Credits

Authors
Andy Kirk

Reviewers
Alberto Cairo
Ben Jones
Santiago Ortiz
Jerome Cukier

Acquisition Editor
Joanna Finchen

Lead Technical Editor
Shreerang Deshpande

Technical Editor
Dominic Pereira

Project Coordinator
Joel Goveya

Proofreader
Chris Brown

Indexer
Tejal Soni

Graphics
Aditi Gajjar

Production Coordinator
Prachali Bhiwandkar

Cover Work
Prachali Bhiwandkar

About the Author

Andy Kirk is a freelance data visualization design consultant, training provider, and editor of the popular data visualization blog, `visualisingdata.com`.

After graduating from Lancaster University with a B.Sc. (Hons) degree in Operational Research, he spent over a decade at a number of the UK's largest organizations in a variety of business analysis and information management roles.

Late 2006 provided Andy with a career-changing "eureka" moment through the serendipitous discovery of data visualization and he has passionately pursued this subject ever since, completing an M.A. (with Distinction) at the University of Leeds along the way.

In February 2010, he launched `visualisingdata.com` with a mission to provide readers with inspiring insights into the contemporary techniques, resources, applications, and best practices around this increasingly popular field. His design consultancy work and training courses extend this ambition, helping organizations of all shapes, sizes, and industries to enhance the analysis and communication of their data to maximize impact.

This book aims to pass on some of the expertise Andy has built up over these years to provide readers with an informative and helpful guide to succeeding in the challenging but exciting world of data visualization design.

Thanks go to my family and friends, but especially to my wonderful wife, Ellie, for her unwavering support, patience, and guidance.

Data Visualization: a successful design process

A structured design approach to equip you with the
knowledge of how to successfully accomplish any
data visualization challenge efficiently and effectively

Andy Kirk

[PACKT]
PUBLISHING

BIRMINGHAM - MUMBAI

Data Visualization: a successful design process

First published: December 2012

Production Reference: 1191212

Published by Packt Publishing Ltd.
Livery Place
35 Livery Street
Birmingham B3 2PB, UK.

ISBN 978-1-84969-346-2

www.packtpub.com

Cover Image by Duraid Fatouhi (duraidfatouhi@yahoo.com)

About the Reviewers

Alberto Cairo has taught infographics and data visualization at the University of Miami since January 2012. He is the author of the book *The Functional Art: An Introduction to Information Graphics and Visualization* (*Peachpit/Pearson*, 2012, http://www.thefunctionalart.com). He has been director of infographics at El Mundo online, Spain (2000-2005), professor of infographics and visualization at the University of North Carolina-Chapel Hill (2005-2009), and director of infographics and multimedia at Época magazine, Brazil (2010-2011). In the past decade, he has consulted with media organizations and educational institutions in nearly 20 countries.

Ben Jones is founder of Data Remixed, a website dedicated to exploring and sharing data analysis and data visualization in an engaging way. Ben has a mechanical engineering and business (entrepreneurship) background, and has spent time as a process improvement expert and trainer in Corporate America. Ben specializes in creating interactive data visualizations with Tableau software, and has won a number of Tableau data visualization competitions. This is Ben's first contribution to a book on the subject of data visualization.

I'd like to thank Andy Kirk for selecting me to contribute as a technical reviewer of this book, and my wife Sarah for all the support she gives me in pursuing my passion of the field of data visualization. I'd also like to thank my fellow technical reviewers, from whom I have learned a great deal over the course of the creation of this book.

Santiago Ortiz invents and develops highly innovative and interactive projects for the Web, using self-built frameworks in JavaScript, HTML5, and ActionScript.

He has over more than 10 years of experience working on interactive visualization projects. In 2005, he co-founded Bestiario (`http://bestiario.org`), the first European company specializing in information visualization. Currently, he freelances in the U.S.A. and Europe.

He has presented at events such as VISWEEK, FutureEverything, VizEurope, O'Reilly STRATA, SocialMediaWeek, NYViz, OFFF, and ARS ELECTRONICA.

His projects have been featured in blogs such as ReadWriteWeb, FlowingData, O'REILLY radar, Fast CoDesign, Gizmodo, and The Guardian datablog.

Jerome Cukier is a highly respected Paris-based data visualization consultant with many years of experience as a data analyst and coordinator of data visualization initiatives at the OECD. Jerome specializes in the creation and design

of data visualizations, data analytics, and gamification. His broad portfolio of work is regularly profiled on the leading visualization and design websites and collated on his own site at `http://www.jeromecukier.net`.

www.PacktPub.com

Support files, eBooks, discount offers and more

You might want to visit www.PacktPub.com for support files and downloads related to your book.

Did you know that Packt offers eBook versions of every book published, with PDF and ePub files available? You can upgrade to the eBook version at www.PacktPub.com and as a print book customer, you are entitled to a discount on the eBook copy. Get in touch with us at service@packtpub.com for more details.

At www.PacktPub.com, you can also read a collection of free technical articles, sign up for a range of free newsletters and receive exclusive discounts and offers on Packt books and eBooks.

http://PacktLib.PacktPub.com

Do you need instant solutions to your IT questions? PacktLib is Packt's online digital book library. Here, you can access, read and search across Packt's entire library of books.

Why Subscribe?

- Fully searchable across every book published by Packt
- Copy and paste, print and bookmark content
- On demand and accessible via web browser

Free Access for Packt account holders

If you have an account with Packt at www.PacktPub.com, you can use this to access PacktLib today and view nine entirely free books. Simply use your login credentials for immediate access.

Table of Contents

Preface

Welcome to the craft of data visualization—a multidisciplinary recipe of art, science, math, technology, and many other interesting ingredients. Not too long ago we might have associated charting or graphing data as a specialist or fringe activity—it was something that scientists, engineers, and statisticians did.

Nowadays, the analysis and presentation of data is a mainstream pursuit. Yet, very few of us have been taught how to do these types of tasks well. Taste and instinct normally prove to be reliable guiding principles, but they aren't sufficient alone to effectively and efficiently navigate through all the different challenges we face and the choices we have to make.

This book offers a handy strategy guide to help you approach your data visualization work with greater know-how and increased confidence. It is a practical book structured around a proven methodology that will equip you with the knowledge, skills, and resources required to make sense of data, to find stories, and to tell stories from your data.

It will provide you with a comprehensive framework of concerns, presenting step-by-step all the things you have to think about, advising you when to think about them and guiding you through how to decide what to do about them.

Once you have worked through this book, you will be able to tackle any project—big, small, simple, complex, individual, collaborative, one-off, or regular—with an assurance that you have all the tactics and guidance needed to deliver the best results possible.

What this book covers

Chapter 1, The Context of Data Visualization, provides an introduction to the subject, its value and relevance today, including some foundation understanding around the theoretical and practical basis of data visualization. This chapter introduces the data visualization methodology and the step-by-step approach recommended to achieve effective and efficient designs. We finish off with a discussion about some of the fundamental design objectives that provide a valuable reference for the suitability of the choices we subsequently make.

Chapter 2, Setting the Purpose and Identifying Key Factors, launches the methodology with the first stage, which is concerned with the vital task of identifying the purpose of your visualization—what is its reason for existing and what is its intended effect? We will look closely at the definition of a visualization's function and its tone in order to shape our design decision-making at the earliest possible opportunity. To complete this scoping stage we will identify and assess the impact of other key factors that will have an effect on your project. We will pay particularly close attention to the skills, knowledge, and general capabilities that are necessary to accomplish an effective visualization solution.

Chapter 3, Demonstrating Editorial Focus and Learning About Your Data, looks at the intertwining issues of the data we're working with and the stories we aim to extract and present. We will look at the importance of demonstrating editorial focus around what it is we are trying to say and then work through the most time-consuming aspect of any data visualization project—the preparation of the data. To further cement the learning in this chapter, we will look at an example of how we use visualization methods to find and tell stories.

Chapter 4, Conceiving and Reasoning Visualization Design Options, takes us beyond the vital preparatory and scoping stages of the methodology and towards the design issues involved in establishing an effective visualization solution. This is arguably the focal point of the book as we look to identify all the design options we have to consider and what choices to make. We will work through this stage by forensically analyzing the anatomy of a visualization design, separating our challenge into the complementary dimensions of the representation and presentation of data.

Chapter 5, Taxonomy of Data Visualization Methods, goes hand-in-hand with the previous chapter as it explores the taxonomy of data visualization methods as defined by the primary communication purpose. Within this chapter we will see an organized collection of some of the most common chart types and graphical methods being used that will provide you with a gallery of ideas to apply to your own projects.

Chapter 6, Constructing and Evaluating Your Design Solution, concludes the methodology by focusing on the final tasks involved in constructing your solution. This chapter will outline a selection of the most common and useful software applications and programming environments. It will present some of the key issues to think about when testing, finishing, and launching a design solution as well as the important matter of evaluating the success of your project post-launch. Finally, the book comes to a close by sharing some of the best ways for you to continue to learn, develop, and refine your data visualization design skills.

What you need for this book

As with most skills in life that are worth pursuing, to become a capable data visualization practitioner takes time, patience, and practice.

You don't need to be a gifted polymath to get the most out of this book, but ideally you should have reasonable computer skills (software and programming), have a good basis in mathematics, and statistics in particular, and have a good design instinct.

There are many other facets that will, of course, be advantageous but the most important trait is just having a natural creativity and curiosity to use data as a means of unlocking insights and communicating stories. These will be key to getting the maximum benefit from this text.

You cannot become skilled by reading this book alone, so you need to have a realistic perspective about the journey you are taking and the distance you have made already. However, by applying the techniques presented, then learning and developing from your experiences, you will enjoy a continued and successful process of improvement.

Who this book is for

Regardless of whether you are an experienced visualizer or a rookie just starting out, this book should prove useful for anyone who is serious about wanting to optimize his or her design approach.

The intention of this book is to be something for everyone—you might be coming into data visualization as a designer and want to bolster your data skills, you might be strong analytically but want inspiration for the design side of things, you might have a great nose for a story but don't quite possess the means for handling or executing a data-driven design.

Some of you may never actually fulfill the role of a designer and might have other interests in learning about data visualization. You may be commissioning work or coordinating a project team and want to know how to successfully handle and evaluate a design process.

Hopefully, it will inform and inspire all who wish to get involved in data visualization design work regardless of role or background.

Conventions

In this book, you will find a number of styles of text that distinguish between different kinds of information. Here are some examples of these styles, and an explanation of their meaning.

New terms and **important words** are shown in bold. Words that you see on the screen, in menus or dialog boxes for example, appear in the text like this:

"**Explanatory** data visualization is about conveying information to a reader in a way that is based around a specific and focused narrative."

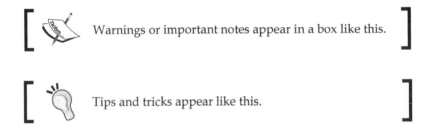

> Warnings or important notes appear in a box like this.

> Tips and tricks appear like this.

Reader feedback

Feedback from our readers is always welcome. Let us know what you think about this book—what you liked or may have disliked. Reader feedback is important for us to develop titles that you really get the most out of.

To send us general feedback, simply send an e-mail to feedback@packtpub.com, and mention the book title via the subject of your message.

If there is a topic that you have expertise in and you are interested in either writing or contributing to a book, see our author guide on www.packtpub.com/authors.

Customer support

Now that you are the proud owner of a Packt book, we have a number of things to help you to get the most from your purchase.

Errata

Although we have taken every care to ensure the accuracy of our content, mistakes do happen. If you find a mistake in one of our books—maybe a mistake in the text or the code—we would be grateful if you would report this to us. By doing so, you can save other readers from frustration and help us improve subsequent versions of this book. If you find any errata, please report them by visiting http://www.packtpub.com/support, selecting your book, clicking on the **errata submission form** link, and entering the details of your errata. Once your errata are verified, your submission will be accepted and the errata will be uploaded on our website, or added to any list of existing errata, under the Errata section of that title. Any existing errata can be viewed by selecting your title from http://www.packtpub.com/support.

Piracy

Piracy of copyright material on the Internet is an ongoing problem across all media. At Packt, we take the protection of our copyright and licenses very seriously. If you come across any illegal copies of our works, in any form, on the Internet, please provide us with the location address or website name immediately so that we can pursue a remedy.

Please contact us at copyright@packtpub.com with a link to the suspected pirated material.

We appreciate your help in protecting our authors, and our ability to bring you valuable content.

Questions

You can contact us at questions@packtpub.com if you are having a problem with any aspect of the book, and we will do our best to address it.

1
The Context of
Data Visualization

This opening chapter provides an introduction to the subject of data visualization and the intention behind this book.

We start things off with some context about the subject. This will briefly explain why there is such an appetite for data visualization and why it is so relevant in the modern age against the backdrop of enhanced technology, increasing capture and availability of data, and the desire for innovative forms of communication.

After this introduction, we then look at the theoretical basis of data visualization, specifically the importance of understanding visual perception. To help establish a term of reference for the rest of the book, we'll then consider a proposed definition for this subject.

Next, we introduce the data visualization methodology, a recommended approach that forms the core of this book, and discuss its role in supporting an effective and efficient design process.

Finally, we consider some of the fundamental data visualization design objectives. These provide a useful framework for evaluating the suitability of the choices we make along the journey towards an accomplished design solution.

Exploiting the digital age

The following is a quotation from Hal Varian, Google's chief economist (http://www.mckinseyquarterly.com/Hal_Varian_on_how_the_Web_challenges_managers_2286):

> *The ability to take data — to be able to understand it, to process it, to extract value from it, to visualize it, to communicate it — that's going to be a hugely important skill in the next decades.*

Data visualization is not new; the visual communication of data has been around in various forms for hundreds and arguably thousands of years. Popular methods that still dominate the boardrooms of corporations across the land — the line, bar, and pie charts — originate from the eighteenth century.

What *is* new is the contemporary appetite for and interest in a subject that has emerged from the fringes and into mainstream consciousness over the past decade.

Catalyzed by powerful new technological capabilities as well as a cultural shift towards greater transparency and accessibility of data, the field has experienced a rapid growth in enthusiastic participation.

Where once the practice of this discipline would have been the preserve of specialist statisticians, engineers, and academics, the globalized field that exists today is a very active, informed, inclusive, and innovative community of practitioners pushing the craft forward in fascinating directions. The following image shows a screenshot of the OECD 'Better Life Index', comparing well-being across different countries. This is just one recent example of an extremely successful visual tool emerging from this field.

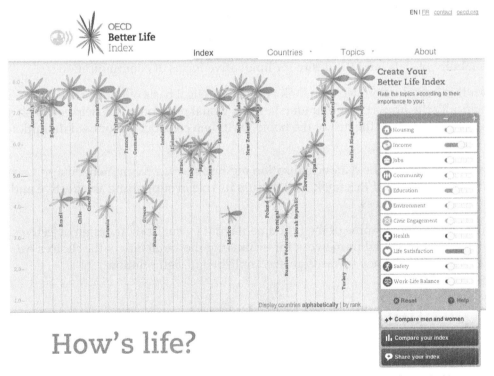

Image from "OECD Better Life Index" (http://oecdbetterlifeindex.org), created by Moritz Stefaner (htpp://moritz.stefaner.eu) in collaboration with Raureif GmbH (http://raureif.net)

Data visualization is the multi-talented, boundary-spanning trendy kid that has seen many esteemed people over the past few years, such as Hal Varian, forecasting this as one of the next big things.

Anyone considering data visualization as a passing fad or just another vacuous buzzword is short-sighted; the need to make sense of and communicate data to others will surely only increase in relevance. However, as it evolves from the *next* big thing to the *current* big thing, the field is at an important stage of its diffusion and maturity. Expectancy has been heightened and it does have a certain amount to prove; something concrete to deliver beyond just experimentation and constant innovation.

It is an especially important discipline with a strong role to play in this modern age. To help frame this, let's first look at the data side of things.

Take a minute to imagine your data footprint over the past 24 hours; that is, the activities you have been involved in or the actions you have taken that will have resulted in data being created and captured.

You've probably included things such as buying something in a shop, switching on a light, putting some fuel in your car, or watching a TV program: the list can go on and on.

Almost everything we do involves a digital consequence; our lives are constantly being recorded and quantified. That sounds a bit scary and probably a little too close for comfort to Orwell's dystopian vision. Yet, for those of us with an analytical curiosity, the amount of data being recorded creates exciting new opportunities to make and share discoveries about the world we live in.

Thanks to incredible advancements and pervasive access to powerful technologies we are capturing, creating, and mobilizing unbelievable amounts of data at an unbelievable rate. Indeed, such is the exponential growth in digital information, in the last two years alone, humanity has created more data than had ever previously been amassed (`http://www.emc.com/leadership/programs/digital-universe.htm`).

Data is now rightly seen as an invaluable asset, something that can genuinely help change the world for the better or potentially create a competitive goldmine, depending on your perspective. "Data is the new oil", first voiced in 2006 and attributed to Clive Humby of Dunnhumby, is a term gaining traction today. Corporations, government bodies, and scientists, to name but a few, are realizing the challenges and, moreover, opportunities that exist with effective utilization of the extraordinary volumes, large varieties, and great velocity of data they govern.

However, to unlock the potential contained within these deep wells of ones and zeros requires the application of techniques to explore and convey the key insights.

Flipping to the opposite side of the data experience, we also identify ourselves as consumers of data. As you would expect, given the volume of captured data, never before in our history have we been faced with the prospect of having to process and digest so much.

Through newspapers, magazines, advertising, the Web, text messaging, social media, and e-mail, our eyes and brains are being relentlessly bombarded by information. In a typical day, it is said we can expect to consume about 100,000 words (`http://hmi. ucsd.edu/howmuchinfo_research_report_consum.php`), which is an astonishing quantity of signals for us to have to make sense of.

Unquestionably, a majority of this visual onslaught flies past us without consequence. We see much of it as noise and we zone out as a way of coping with the overload and saturation of things to think and care about.

What this shows is the necessity to be more effective and efficient in how data is communicated. It needs to be portrayed in ways that help to get our messages across in both an engaging and informative way.

If data is the oil, then data visualization is the engine that facilitates its true value and that is why it is such a relevant discipline for exploiting our digital age.

Visualization as a discovery tool

One of the most compelling arguments for the value of data visualization is expressed in this quote from John W Tukey (*Exploratory Data Analysis*).

> *The greatest value of a picture is when it forces us to notice what we never expected to see.*

Through visualization, we are seeking to portray data in ways that allow us to see it in a new light, to visually observe patterns, exceptions, and the possible stories that sit behind its raw state. This is about considering visualization as a tool for discovery.

A well known demonstration that supports this notion was developed by noted statistician Francis Anscombe (incidentally, brother-in-law to Tukey) in the 1970s. He compiled an experiment involving four sets of data, each exhibiting almost identical statistical properties including mean, variance, and correlation. This was known as "Anscombe's quartet".

x1	y1		x2	y2		x3	y3		x4	y4
10	8.04		10	9.14		10	7.46		8	6.58
8	6.95		8	8.14		8	6.77		8	5.76
13	7.58		13	8.74		13	12.74		8	7.71
9	8.81		9	8.77		9	7.11		8	8.84
11	8.33		11	9.26		11	7.81		8	8.47
14	9.96		14	8.1		14	8.84		8	7.04
6	7.24		6	6.13		6	6.08		8	5.25
4	4.26		4	3.1		4	5.39		19	12.5
12	10.84		12	9.13		12	8.15		8	5.56
7	4.82		7	7.26		7	6.42		8	7.91
5	5.68		5	4.74		5	5.73		8	6.89

Sample data sets recreated from Anscombe, Francis J. (1973) Graphs in statistical analysis. American Statistician, 27, 17–21

Ask yourself, what can you *see* in these sets of data? Do any patterns or trends jump out? Perhaps the sequence of eights in the fourth set? Otherwise there's nothing much of interest evident.

So what if we now visualize this data, what can we see then?

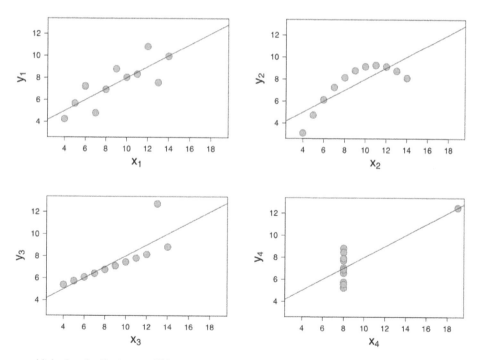

Image published under the terms of "Creative Commons Attribution-Share Alike", source: http://commons. wikimedia.org/wiki/File:Anscombe%27s_quartet_3.svg

Through the previous graphical display, we can immediately see the prominent patterns created by the relationships between the X and Y values across the four sets of data as follows:

- the general tendency about a trend line in **X1, Y1**
- the curvature pattern of **X2, Y2**
- the strong linear pattern with single outlier in **X3, Y3**
- the similarly strong linear pattern with an outlier for **X4, Y4**

The intention and value of Anscombe's experiment was to demonstrate the importance of presenting data graphically. Rather than just describing a dataset based on a selection of some of its key statistical properties alone, to make proper sense of data, and avoid forming false conclusions we need to also employ visualization techniques.

It is much easier to discover and confirm the presence (or even absence) of patterns, relationships, and physical characteristics (such as outliers) through a visual display, reinforcing the essence of Tukey's quote about the value of pictures.

Data visualization is about a discovery process, enabling the reader to move from just looking at data to actually seeing it. This is a subtle but important distinction.

The bedrock of visualization knowledge

Data visualization is not easy. Let's make that clear from the start. It should be genuinely viewed as a craft. It is a unique convergence of many different skills and requires a great deal of practice and experience, which clearly demands time and patience.

Above all, it requires a deep and broad knowledge across several traditionally discrete subjects, including cognitive science, statistics, graphic design, cartography, and computer science.

This multi-disciplinary recipe unquestionably makes it a challenging subject to master but equally provides an exciting proposition for many. This is evidenced by the field's popular participation, drawing people from many diverse backgrounds.

If we look at this subject convergence at a more summary level, data visualization could be described as an intersection of art and science. This combination of creative and scientific perspectives represents a delicate mixture. Achieving an appropriate balance between these contrasting ingredients is one of the fundamental factors that will determine the success or failure of a designer's work.

The *art* side of the field refers to the scope for unleashing design flair and encouraging innovation, where you strive to design communications that appeal on an aesthetic level and then survive in the mind on an emotional one. Some of the modern-day creative output from across the field is extraordinary and we'll see a few examples of this throughout the chapters ahead.

The *science* behind visualization comes in many shapes. I've already mentioned the presence of computer science, mathematics, and statistics, but one of the key foundations of the subject comes through an understanding of cognitive science and in particular the study of visual perception. This concerns how the functions of the eye and the brain work together to process information as visual signals.

One of the other most influential founding studies about visual perception emerged from the Gestalt School of Psychology in the early 1900s, specifically in the shape of the Laws of Perceptual Organization (http://www.interaction-design.org/encyclopedia/data_visualization_for_human_perception.html).

These laws provide an organized understanding about the different ways our eyes and brain inherently and automatically form a global sense of patterns based on the arrangement and physical attributes of individual elements.

Here, we can see two visual examples of Gestalt Laws.

On the left-hand side is a demonstration of the "Law of Similarity". This shows a series of rows with differently shaded circles. When we see this our visual processes instantly determine that the similarly shaded circles are related and part of a group that is separate and different to the non-shaded rows. We don't need to think about this and wait to form such a conclusion; it is a preattentive reaction.

Images republished from the freely licensed media file repository Wikimedia Commons, source: http://en.wikipedia.org/wiki/File:Gestalt_similarity.svg and http://en.wikipedia.org/wiki/File:Gestalt_proximity.svg

On the right-hand side is a demonstration of the "Law of Proximity". The arrangement of closely packed-together pairs of columns means we assume these to be related and distinct from the other pairings. We don't really view this display as six columns, rather we view them as three clusters or sets.

At the root of visual perception knowledge is the understanding that our visual functions are extremely fast and efficient processes whereas our cognitive processes, the act of thinking, is much slower and less efficient. How we exploit these attributes in visualization has a significant impact on how effectively the design will aid interpretation.

Consider the following examples, both portraying analysis of the placement of penalties taken by soccer players.

When we look at the first image, the clarity of the display allows us to instantly identify the football symbols, their position, and their classifying color. We don't need to think about how to interpret it, we just do. Our thoughts, instead, are focused on the consequence of this information: what do these patterns and insights mean to us? If you're a goalkeeper, you'll be learning that, in general, the penalty taker tends to place their shots to the right of the goal.

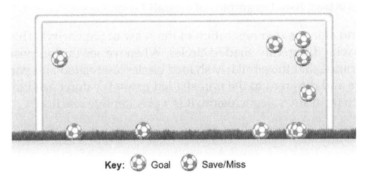

Key: 🔵 Goal ⚪ Save/Miss

Image republished under the terms of "fair use", source: http://www.facebook.com/castrolfootball

By contrast, this second display's attempt to portray the same type of data presentation causes significant visual clutter and confusion. Rather than using a simple and relatively blank image like the previous one, this display includes strong colors and imagery in the background. The result is that our eyes and brain have to work much harder to spot the footballs and their colors because the data layer has to compete for attention with the background imagery. We are therefore unable to rely on the capabilities of our preattentive visual perception (determined by the Law of Similarity) because we cannot easily perceive the shapes and their attributes representing the data. This delays our interpretative processes considerably and undermines the effectiveness and efficiency of the communication exchange.

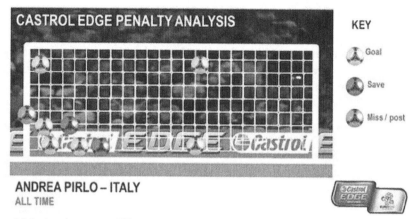

Image republished under terms of "fair use", source: http://www.mirror.co.uk/sport/football/
euro-2012-where-italy-will-place-their-penalties-907506

This is just a single, simple example but it does reveal the significance of understanding and obeying visual perception laws when portraying our data.

When we design a visualization, we need to take advantage of the strengths of the visual function and avoid the disadvantages of the cognitive functions. We need to minimize the amount of thinking or "working out" that goes into reading and interpreting data and simply let the eyes do their efficient and effective job.

Through the pioneering studies and development of theories acquired and refined over many years by the Gestalt School of Psychology as well as influential academics and theorists like Jacques Bertin, Francis Anscombe, John W Tukey, Jock McKinlay, and William Cleveland, we now have a greater understanding of how to achieve effective and efficient visualization design.

There is still a great amount of empirical evidence to gather, studies to conduct, and firm answers to unearth, but the wealth of knowledge available to us is a significant help to remove an undue amount of instinct in our design work.

Defining data visualization

It is important now to consider a definition of data visualization. To do this, we first need to consider the main agents involved in the exchange of information; namely, the messenger, the receiver, and the message. The relationship between these three is clearly very important, as this illustration explains:

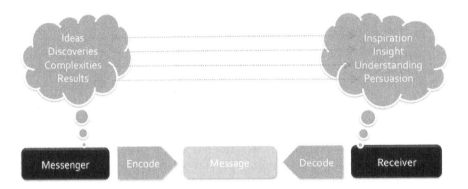

On one side we have a messenger looking to impart results, analysis, and stories. This is the designer. On the other side, you have the receiver of the message. These are the readers or the users of your visualization. The message in the middle is the channel of communication. In our case this is the data visualization; a chart, an online interactive, a touch screen installation, or maybe an infographic in a newspaper. This is the form through which we communicate to the receiver.

The task for you as the designer is to put yourself in the shoes of the reader. Try to imagine, anticipate, and determine what they are going to be seeking from your message. What stories are they seeking? Is it just to learn something new or are they looking for persuasion, something with more emotional impact? This type of appreciation is what fundamentally shapes the best practices in visualization design: considering and respecting the needs of the reader.

The important point is this: to ensure that our message is conveyed in the most effective and efficient form, one that will serve the requirements of the receiver, we need to make sure we design (or "encode") our message in a way that actively exploits how the receiver will most effectively interpret (or "decode") the message through their visual perception capabilities.

From this illustration we can form the following definition to clarify, at this early stage, what we mean by data visualization:

The representation and presentation of data that exploits our visual perception abilities in order to amplify cognition.

Let's take a closer look at the key elements of this definition to clarify its meaning; these are as follows:

- The **representation** of data is the way you decide to depict data through a choice of physical forms. Whether it is via a line, a bar, a circle, or any other visual variable, you are taking data as the raw material and creating a representation to best portray its attributes. We will cover this aspect of design much more in *Chapter 4, Conceiving and Reasoning Visualization Design Options* and *Chapter 5, Taxonomy of Data Visualization Methods*.

- The **presentation** of data goes beyond the representation of data and concerns how you integrate your data representation into the overall communicated work, including the choice of colors, annotations, and interactive features. Similarly, this will be covered in depth in *Chapter 4, Conceiving and Reasoning Visualization Design Options*.

- Exploiting our **visual perception abilities** relates to the scientific understanding of how our eyes and brains process information most effectively, as we've just discussed. This is about harnessing our abilities with spatial reasoning, pattern recognition, and big-picture thinking.

- **Amplify cognition** is about maximizing how efficiently and effectively we are able to process the information into thoughts, insights, and knowledge. Ultimately, the objective of data visualization should be to make a reader or users feel like they have become better informed about a subject.

The definition that I've put forward here is not dissimilar to the many others articulated by authors, academics, and designers down the years. It is not intended to offer a paradigm shift in our understanding of what this is all about. Rather, it represents a personal perspective of the discipline influenced by many years of experience teaching, practicing, and constantly studying the subject.

The fact that data visualization is such a dynamic and evolving field, with this unique conjunction of art and science shaping its practice, means that a single, perfect, and universally-agreed definition is always going to be difficult to construct. However, this proposed definition should at least help you develop an appreciation of the boundaries of data visualization and recognize when something evolves into a different form of creative output.

Visualization skills for the masses

The following is a quote from Stephen Few from his book *Show Me the Numbers*:

> *"The skills required for most effectively displaying information are not intuitive and rely largely on principles that must be learned."*

More and more of us are becoming responsible for the analysis, presentation, and interpretation of data. This naturally reflects the explosion in access to data and the value attributed to potential insights that are contained.

As I've already stated, where once this was typically a specialist role, nowadays the responsibility for dealing with data has crept into most professional duties. This has been accelerated by the ubiquitous availability of a range of accessible productivity tools to handle and analyze data.

This means visualization has become both a problem and an opportunity for the masses, which makes the importance and dissemination of effective practice a key imperative.

The quote from Stephen Few will resonate with many of you reading this. If you were to ask yourself "Why do I design visualizations in the way I do?", what would be your answer? Think about any chart or graphic you produce to communicate information to others. How do you design it? What factors do you take into account? Perhaps your response would fall in to one or more of the following:

- You have a certain design style based on personal taste
- You just play around until something emerges that you instinctively like the look of
- You trust software defaults and don't go beyond that in terms of modifying the design
- You have limited software capabilities, so you don't know how to modify a design
- You just do as the boss tells you — "can you do me some fancy charts?"

For many people, the idea of a conscious data visualization design technique is quite new. The absence of any formal coaching, at almost any level of education, in the techniques of visualization means until you become aware of the subject, you have probably never even thought about your visualization design approach.

Before discovering this subject, my own approach to presenting data was certainly not informed by any training or prior knowledge. I'd never even thought about it. Taste and gut-feel were my guiding principles alongside a perceived need to show off technical competencies in tools like Excel. Indeed, I'd like to take this opportunity to apologize for much of my graphical output between 1995 and 2005 where striking gradients and "impressive" 3D were commonplace. The thing is, as I've just said, I didn't realize there was a better way; it simply wasn't on my radar.

In some respects, the reliance on instinct, playing about with solutions that seem to work fine for us, can suffice for most of our needs. However, these days, you often hear the desire being expressed to move beyond devices like the bar chart and find different creative ways to communicate data.

While it is a perfectly understandable desire, just aiming for something different (or even worse, something "cool") is not a good enough motive in itself.

If we want to optimize the way we approach a data visualization design, whether it be a small, simple chart or a complicated interactive graphic, we need to be better equipped with the necessary knowledge and appreciation of the many design and analytical decisions we need to make.

As suggested previously, instinct and taste have got us so far but to move on to a whole new level of effectiveness, we need to understand the key design concepts and learn about the creative process. This is where the importance of a methodology comes in.

The data visualization methodology

The design methodology described in this book is intended to be portable to any visualization challenge. It presents a sequence of important analytical and design tasks and decisions that need to be handled effectively.

As any fellow student of Operational Research (the "Science of Better") will testify, through planning and preparation, and the development and deployment of strategy, complex problems can be overcome with greater efficiency, effectiveness, and elegance. Data visualization is no different.

Adopting this methodology is about recognizing the key stages, considerations, and tactics that will help you navigate smoothly through your visualization project.

Remember, though, design is rarely a neat, linear process and indeed some of the stages may occasionally switch in sequence and require iteration. It is natural that new factors can emerge at any stage and influence alternative solutions, so it is important to be open-minded and flexible. Things might need to be revisited, decisions reversed, and directions changed. What we are trying to do, where possible, is find the best path through the minefield of design choices.

Some may feel uncomfortable at the prospect of following a process to undertake what is fundamentally an iterative, creative design process. But I would argue everyone should find value from working in a more organized and sequenced way especially if it helps to reduce inefficiency and wasted resource.

The design challenges involved in data visualization are predominantly technology related; the creation and execution of a visualization design will typically require the assistance of a variety of applications and programs. However, the focus of this methodology is intended to be technology-neutral, placing an emphasis on the concepting, reasoning, and decision-making.

The variety, evolution, and generally fragmented nature of software in this field (there is no single tool that can do everything) highlights the extra importance of reasoned decision-making, regardless of the richness and power individual solutions can offer.

Another key point to remark on is to emphasize, if it wasn't already clear, that data visualization is not an exact science. There is rarely, if ever, a single right answer or single best solution. It is much more about using heuristic methods to determine the most satisfactory solutions.

On that note, the content of the methodology intentionally avoids any sense of dogmatic instruction, preferring to focus on guidelines over explicit rules; sometimes an ounce of chaos, a certain license to experiment, a leaning on instinct, and a sense of randomness can spark greater creativity and serendipitous discovery.

The methodology is intended to be adopted flexibly, based on your own judgment and discretion, by simply laying out all the important things you need to take into account and proposing some potential solutions for different scenarios.

Finally, as I stressed with my definition of the subject earlier, I'm not suggesting this is a ground-breaking new take on the creative process. It is merely a personal interpretation based on experience and also exposure to the many brilliant people out there who share their own design narratives. It is, though, consistent with how most established observers of the subject would recommend you undertake this task. Moreover, it is an approach that I fundamentally believe works and it has genuinely helped me improve my own work since I've adopted it more deliberately, allowing me to cut through projects with the efficiency and elegance I've always yearned for.

Visualization design objectives

Before we launch in to the first stages of the methodology in *Chapter 2, Setting the Purpose and Identifying Key Factors*, it is important to acknowledge a handful of key, overriding design objectives that should provide you with a framework to test your progress and the suitability of your design decisions.

Whereas the methodology will introduce a number of key thoughts and decisions at each stage of the process, these objectives transcend any individual step and highlight the intricate issues you have to handle throughout your process.

The key objectives are as follows:

Strive for form and function

The following is a quote from Frank Lloyd Wright:

> *"Form follows function – that has been misunderstood. Form and function should be one, joined in a spiritual union."*

The first objective brings us immediately face-to-face with the age-old debate of form versus function or style over substance. As Frank Lloyd Wright proposed, all the way back in 1908, these are aspects of design that should be combined and brought together in harmony, not at the sacrifice of one or the other. There's room and a need for both.

It is a very difficult balancing act to achieve, as I've already alluded to in the discussion about art and science, but our aim should be to hit that sweet-spot where something is aesthetically inviting and functionally effective.

The designer and author Don Norman (`http://www.jnd.org/dn.mss/emotion_design.html`) talks about how we're more tolerant about things that are attractive and more likely to want them to perform well. Indeed, there is a school of thought that suggests how we think cannot be separated from how we feel.

Norman goes on to describe how well-executed aesthetics can naturally create favorable emotional and mental responses, but emotional affection can also come from the experience of good usability and the accomplishment of insight. Fundamentally, attractive form enhances function and the function portrays beauty through its effect.

Throughout this book, we will see examples of designs that have succeeded in creating elegance in form and in function. The following image is taken from an animated wind map developed by Fernanda Viégas and Martin Wattenberg. It is a beautiful piece of work, exceptionally well designed and executed but it also serves its purpose as a way of informing users about the wind patterns, strength, and directions occurring across the United States. This is form and function in spiritual union:

Image from "Wind Map" (http://hint.fm/wind/) created by Fernanda Viégas and Martin Wattenberg

The general advice, especially for beginners, is to initially focus on securing the functional aspects of your visualization. First, try to achieve the foundation of something that informs—that functions—before exploring the ways of enhancing its form. The simplest analogy would be build the house before decorating it, but I wouldn't want to create too much separation between the two as they are often intrinsically linked. Over time, you will be much more confident and capable of synthesizing the two demands in harmony. We shall discuss this in more depth in *Chapter 4, Conceiving and Reasoning Visualization Design Options*.

Justifying the selection of everything we do

The following is a quote from Amanda Cox (http://vimeo.com/29391942), who works as a graphics editor at the New York Times:

> *"We're so busy thinking about if we can do things, we forget to consider whether we should."*

In many ways, the central idea behind the methodology is encouraging you to determine that everything you do is thoroughly planned, understood, and reasoned.

This particular objective is about recognizing and responding to the scoping information that you will gather at the start of the methodology, to ensure that everything undertaken thereafter serves the purpose of our work and the needs of the audience.

Here, we should consider the idea of deliberate design, which means that the inclusion, exclusion, and execution of every single mark, characteristic, and design feature is done for a reason.

When we reach the stage of designing, concepting, and construction, you should be prepared to challenge everything; the use of a shape, the selection of a color pallet, the position of a label, or the use of an interaction.

In this next example, when displaying a section of a tree-hierarchy work by data illustrator, Stefanie Posavec, every visible property presented is used to communicate data, whether it be the use of color, the arc lengths of the petals, the position and sequence of stems; nothing is redundant and everything is deliberate.

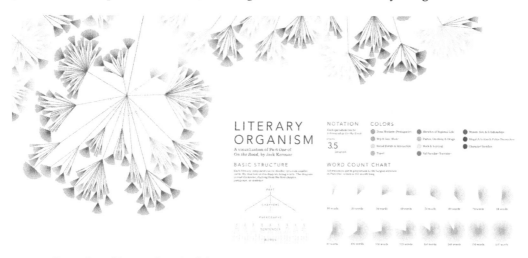

Image from "Literary Organism" (http://itsbeenreal.co.uk/index.php?/wwwords/literary-organism/), created by Stefanie Posavec

It is also important to make sure that any visual property that is included, but does not represent data, such as shading, labels, colors, and axes among other properties, should only be included to aid the process of visual perception, not hinder it.

Furthermore, for interactive and animated visualizations, remember Amanda Cox's quote—"just because you can, doesn't mean you should." Don't succumb to the belief (like I did for many years) of thinking a visualization is a platform solely to showcase your technical competence.

Cluttering visualizations with fancy interactive features is a trap that is easy to fall into and leads to projects that look nice or are impressive technically but fail to serve their intended purpose. Instead, they interfere with the efficiency and effectiveness of the information exchange thus demonstrating a failure to synthesize form and function.

Creating accessibility through intuitive design

The following is a quote from Edward Tufte (`http://adage.com/article/adagestat/edward-tufte-adagestat-q-a/230884/`):

> *"Overload, clutter, and confusion are not attributes of information, they are failures of design."*

When you next happen to be in a town or city center, take a look around you and observe how often people are confused by and struggle with the basic operation of correctly opening and entering doors into a store. Notice how the accessibility and function of a door—the simple act of opening and walking through it—is often impaired through a lack of intuitive design.

The method of opening a door should be straightforward, but often the aesthetics of features such as stylish door handles means we pull when we should push and we push when we should pull. This is a flaw in the intuitiveness and logic of the design, a failure in perceived affordance—it doesn't do what it looks like it should do.

This idea is an important concept to translate into visualization. As we have already outlined, we are trying to exploit the inherent spatial reasoning and pattern recognition functions of visual perception. We don't want people to have to spend unnecessary time thinking about how to use or how to read and interpret something.

When you are creating a visualization, you are integrating visual design with a subject matter's data. The former is the window into the latter, and it is the design and execution of this window that creates the accessibility.

But it is important to create a distinction between accessibility and immediacy. The speed with which you are able to read or interpret a visualization should be determined by the complexity of the subject and the purpose of the project, not by the ineffectiveness of design.

Sometimes subjects are fundamentally simple and the portrayal of the data is straightforward and intuitive. This in turn means the reader's task of interpreting the data should be relatively easy.

On other occasions, a data framework might be more complex. Your challenge will be to respect the complexity and avoid simplifying, diluting, or reducing the essence of this subject. This might mean something is not immediately easy to interpret. Some visualizations will require effort to be put in, forcing the reader to undertake a certain amount of experiential practice in order for the eye and mind to essentially become trained in reading the display.

Think of it being like muscle memory, but for the eye and the brain. We are so used to reading bar charts and line charts that they have become entrenched and programmed into our interpretative toolkit. But when we are faced with something new, something different or seemingly complex, its not always immediately clear how we are supposed to handle it.

In the following example, we see a demonstration of what is quite a complex data framework. This is an image of a legend that was used to explain how to read an innovative visualization to portray three separate indicators of a movie's success. On the left-hand side of the image is the aggregate reviews (the higher the value, the better) and on the right-hand side of the image are both the budget and gross takings (the bigger the gap, the better):

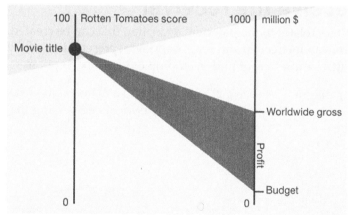

Image from "Spotlight on Profitability" (http://www.szucskrisztina.hu), created by KrisztinaSzucs

It is an unusual representation of data, not something as preprogrammed as the bar or line chart, and so it takes a short while to learn how to read and interpret the resulting shapes formed by the movie data shown across piece. This is absolutely legitimate as an effective approach to visualizing this data so long as the efforts that go into learning how to read it eventually leads the user to understand it.

Take another example, which portrays the key events in a couple of soccer matches showing completed passes (green lines), shots (blue triangles), and goals (red dots) as shown in the following image:

Image from "Umbro World Cup Poster" (`http://www.mikemake.com/Umbro-s-World-Cup-Poster`), created by Michael Deal

Once the reader has mastered the understanding of what each shape and its position means, these displays provide a powerful and rewarding insight in to the key incidents and the general ebb and flow of each game.

In simple terms, so long as you can avoid all the negative characteristics that Edward Tufte mentions at the top of this section, you should succeed in giving people an accessible route in to the data. Make sure that the efforts needed from the reader or user to understand how to use and interpret a visualization are ultimately rewarded with a worthy amount of insight gained.

Never deceive the receiver

Visualization ethics relates to the potential deception that can be created, intentionally or otherwise, from an ineffective and inappropriate representation of data. Sometimes it can be through a simple lack of understanding of visual perception.

In the following diagram, we see a 2D pie chart and a 3D version. When the eye interprets a graphic like this, what it is actually doing is perceiving the proportion of visible pixels:

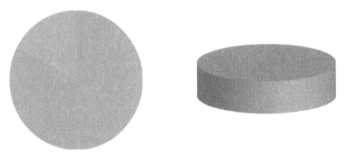

Image from "The Curious Incident of Kevins in Zurich...and other stories" (`http://www.researchobservatories.org.uk/EasysiteWeb/getresource.axd?AssetID=38334`) by Alan Smith.

On the left-hand side of the diagram, we see a blue segment representing 82 percent and an orange segment representing 18 percent. These are the actual values. However, when we introduce a third dimension on the right—incidentally, a dimension which is purely decorative and has no relationship with data values—our eyes are deceived because we are not capable of easily adjusting our interpretation of the values across this isometric projection. With the introduction of the extra dimension and the visible height of the pie itself, we now perceive 91 percent of the visible area as blue and only 9 percent orange. This is clearly a hugely distorted reading of the values.

Another similar example comes from a Wikipedia fundraising campaign from a few years ago and a progress bar depicting the status of their efforts; as shown in the following screenshot:

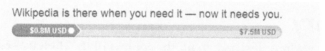

Image published under the terms of "Creative Commons Attribution-Share Alike", source: https://donate.wikimedia.org/

As with the pie chart, for a bar chart we perceive the visible pixels as being representative of the values. The label indicates a total of **$0.8M USD** had been raised (10.7 percent towards target) but if you calculate the actual length of the bar displayed, this occupies 24.6 percent of the overall bar length. Once again, a significant distortion of the truth.

This next example is a demonstration of where aesthetics and style completely hijack a visualization. Here, we have a still showing a 3D bar chart that swooshes impressively onto the screens of those watching soccer on TV in the UK:

But what have we here? There is a yellow **Drawn** bar representing the value **1** and this appears to be more than half the length of a red **Aston Villa** bar representing **4**. How can that be?

The designers of this visual have chosen to include the category labels within the bar's length, thus completely distorting the values being represented. Now, this is possibly one of the least interesting statistics you'll come across, and I'm assured the world will not stop turning as a result of this graphical misdemeanor, but it should demonstrate the pitfalls of decoration and overly stylized design.

Obeying visualization ethics is clearly an objective for any project, but really it is just about basic, good practice, respect for your readers, and attention to detail.

Summary

In this chapter, we have learned about the context of the digital era and the role data visualization can play in helping us make greater sense of the huge volumes of captured data we have access to in today's world.

We have discussed how more and more people are getting involved in activities that require visualization techniques, but the skills required to accomplish this effectively go beyond instinct and require careful learning and practice.

The methodology presented in this book will provide a strategy for designers to develop these techniques through good practice. It will help them navigate through the key decisions that are required throughout the creative process.

Finally, to commence the design thinking, we have learned about some important overriding objectives that should provide a useful assessment of the effectiveness of your visual solution throughout its creation.

In the next chapter, we will commence the data visualization methodology by exploring the first stage of any design challenge: establishing the project's purpose and identifying its inherent key influencing factors.

2
Setting the Purpose and Identifying Key Factors

Chapter 1, Context of Data Visualization, gave us some initial idea about the definition, context, and relevance of data visualization. We learned about the intersection that exists between art and science and outlined the significance of visual perception.

We also discussed the proposed value of a data visualization methodology and, ahead of our design process, learned about some key objectives we need to bear in mind throughout our work.

In this second chapter, we start the design methodology with the vital task of identifying the purpose of your visualization.

Before you undertake any design work, you have to be clear about the motivation behind a project's inception. This involves identifying who it is for and what needs you are trying to fulfill; this has a big influence on the scope of your work.

The second aspect of purpose will take a close look at the intention behind your project and how you define the visualization's function and tone. Once again, we tackle this now so that we can start to shape our decision-making at the earliest possible opportunity.

We then look at identifying and assessing the impact of the additional key factors that will have an effect on your project. This will help you surface all the restrictions, characteristics, and requirements surrounding your project that will determine how you tackle it.

Finally, we will consider in more depth a particularly influential matter: the skills, knowledge, and general capabilities that are necessary to accomplish an effective visualization solution.

Clarifying the purpose of your project

We start at the very beginning. Why are we doing this project? What is its purpose?

These might seem like blunt questions to ask but it is so important to establish this type of clarity before we go much further. You might think: "please can we just get on with it?" and you're probably itching to get on to a computer and start playing with some data, but these initial few stages of this methodology are very deliberately designed to get you in to the habit of this type of reflective or analytical assessment.

If you look at a dictionary definition for "purpose", it will usually say something similar to: reason for existing, intended effect. These two dimensions neatly capture the focus of our definition work at this point in the journey.

The reason for existing

Let's start with the reason for existing. This is about recognizing the trigger behind the project or the origin from where it emerged. This gives us an idea of the scope and context of what we are about to undertake, how much creative control we might have, whether we've been encouraged to follow a particular creative direction and what ideas have already been formed.

A project will typically form in one of following two ways: you've either been asked to do it or you've decided to do something yourself. You might think that's obvious, but these are very different scenarios for working creatively.

If it's the former, a project that's essentially been commissioned, you will have most likely been given the task by a colleague or a manager or from a client you are working with/for. A further source may be something like an invited assignment such as tendering for some work or even entering a design contest.

You will have received or read a brief and possibly had some initial discussions that provided you with an outline of the requirements. You might have some instructions and a general idea of what they are seeking.

From gathering this contextual information, you should have a reasonably clear idea about the background to the project, what you're being asked to do, why you're doing it and who you're doing it for. It may be quite loose and open-ended, in which case you've got a greater range of possibilities, but it's more likely to be quite defined and specific.

You've probably also experienced the pain of hearing some of the initial ideas flying round, as your creative soul dies a slow death in response to requests such as "cool charts" and "make it fancy" or "I want an Edward Tufte style piece".

By contrast, for a project that is self-initiated, things are very different. Maybe, it is a dataset that you've found about a subject that interests you, maybe you've decided to test out some theory or you've been chatting to (or, more truthfully, drinking with) mates and have struck on a particular curiosity that simply must be explored.

This scenario is a completely self-defined, self-determined, and more flexible context than that of a commissioned project. It doesn't involve a client, or a brief, or a set of instructions, or restrictions on scope, timescales, or audience—you've got a blank canvas to follow the scent of what it is that motivated you in the first place.

A very different proposition to a commissioned project and an important distinction that needs to be established.

The intended effect

Whatever the motivation and background for doing the project, you will inevitably start to form a vision in your mind of what you might be about to create, how it might look, and what it might do. This is a natural instinct as you embark on a creative process.

This vision might leap into your mind the minute you start to think around the task, regardless of its origin. You might recall certain influential or inspirational works that you've seen in the past or remember concepts you previously developed that went no further.

It's important to capture these thoughts if they do form. Make sure you keep notes, in your sketchbook, on your tablet, or on a cigarette packet—it doesn't matter where, just do it before you forget. While we don't want to be closed off and commit ourselves to the pursuit of the first thing we think of, these instinctive thoughts could prove valuable later on.

It's from these sparks of creativity that we shift our thoughts to the second dimension of purpose, which is the intended effect of the visualization project. This is a really critical matter, so we're going to take a bit of time to get our heads around it.

Here's why it's so important to be thinking about all this in such depth. Even though it is very early in the process, the decisions, or more accurately, the definitions we form now, will have a strong bearing on the creative direction we pursue. It's not doing anything that can't be reversed or refined further down the line but it is important to establish as much focus and clarity about our intentions now so that we can reduce the complexity of the challenge and the potential variety of the solution.

The choices we make now will also influence the resources (technical and personnel) we might need to deploy, but we shall look more at this towards the end of the chapter.

Remember in *Chapter 1, Context of Data Visualization* (you should do, after all it's only been a few pages) we saw a definition for data visualization that proposed the overall aim as being to "amplify cognition" or, in other words, make someone feel better informed. That was a deliberately equivocal aim because there are many different motivations and reasons for creating a data visualization.

Consider the following sample collection of phrases, which articulate a variety of viable intentions behind creating a visual representation of data:

Lookup   Persuade   Creative technique

Learn/Increase knowledge   Answer questions

Change behaviour   Conduct analysis   Monitor signals

Play with data   Tell story   Trigger questions   Enlighten

Contextualise data   Find patterns/no patterns

Serendipitous discoveries   Familiarise with data

Shape opinion   Emphasize issues   Inspire

Grab attention

Present arguments   Assist decisions

Experimentation

Art/Aesthetic pleasure   Shock/Make an impact

If we look closely at the verbs and the overall language being used, we can start to recognize quite a range of differing effects that might be sought.

For example, a visualization to assist with the monitoring of signals or facilitating a visual lookup of data will be very different from a design that is intended to grab attention or change behavior. Similarly, presenting arguments and telling a story is a very different setting to conducting analysis or 'playing' with data.

What we have here is evidence of different dimensions of intent. Identifying your intended effect means deciding what you're aiming to achieve and how you're going to achieve it.

At the root of this is an appreciation of your target audience, one of the most important considerations we have to take into account. During this initial definition and scoping work, it is crucial to profile your intended readers/users.

These are the people who we are serving and so we need to recognize what type of engagement they are likely to require, for example:

- Is it a boardroom environment with a small collection of senior colleagues who have existing domain subject knowledge?
- Is it a large range of customers, covering all social demographics but potentially representing a captive audience for the subject matter?
- Is it a completely global, undefined audience with no influencing characteristics—in a sense no specific target, just anyone and everyone?
- Is it a one-to-one exchange with a manager?
- Is it an entirely personal engagement between you and data—a desire to learn about and explore data yourself?

Clarity of what and who your target audience is will help shape your intent and from this we'll be able to define two very important dimensions: the function and tone of your visualization project.

Establishing intent – the visualization's function

The intended function of a data visualization concerns the functional experience you create between your design, the data, and the reader/user.

If we revisit the range of phrases presented earlier, it is possible to form three separate clusters or categories of function. While there is always a chance of slight overlap, there will be a significant difference in your design choices depending on whether the function of your visualization is to:

- Convey an **explanatory** portrayal of data to a reader
- Provide an interface to data in order to facilitate visual **exploration**
- Use data as an **exhibition** of self-expression

When the function is to explain

Explanatory data visualization is about conveying information to a reader in a way that is based around a specific and focused narrative. It requires a designer-driven, editorial approach to synthesize the requirements of your target audience with the key insights and most important analytical dimensions you are wishing to convey.

There are many ways in which you can "explain" data. It could be through an information dashboard in a corporate setting, where you are conveying the latest performance figures and highlighting the key issues requiring attention. It might be a graphic in a newspaper, explaining the complexity and severity of the problems around the economic crisis. It could be an animated design to display patterns of population migration over time. It could also be a physical or ambient visualization designed to draw attention to the sugar content of certain drinks.

The end result is typically a visual experience built around a carefully constructed narrative. Your objective as the designer is to create a graphical display, made accessible through intuitive, visual design that clearly portrays the narrative you are seeking to impart.

Here is an example of an explanatory visualization, based on a chart type called a **Sankey** diagram, which portrays analysis of the top ten freshwater-consuming countries and the breakdown of its usage:

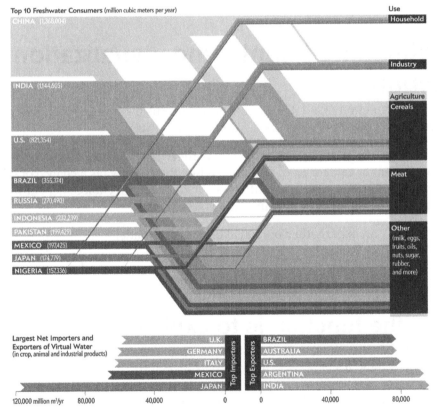

Image from "Top 10 Freshwater Consumers" (http://www.scientificamerican.com/article. cfm?id=water-in-water-out) created by Mark Fischetti and Jen Christiansen, Scientific American, June 2012. Reproduced with permission. Copyright © 2012 Scientific American, Inc. All rights reserved.

Note that explanatory visualizations are not limited to just being static in design. Indeed, some of the most impactive, narrative-driven pieces can be framed within an interactive or animated construction.

When the function is to explore

Exploratory data visualization design is a slightly different matter compared to creating an explanatory piece. Here, we are seeking to facilitate the familiarization and reasoning of data through a range of user-driven experiences. In contrast to explanatory-based functions, exploratory data visualizations lack a specific, single narrative. They are more about visual analysis than just the visual presentation of data.

Exploratory solutions aim to create a tool, providing the user with an interface to visually explore the data. Through this they can seek out personal discoveries, patterns, and relationships, thereby triggering and iterating curiosities. It also opens up the possibility for chance or serendipitous findings caused by forming different combinations of variable displays.

Really, the key feature that differentiates an exploratory piece from an explanatory piece is the amount of work you have to do as a reader to discover insights. For explanatory pieces, the designer should do the hard work and create a clear portrayal of the interesting stories and analysis from a dataset. An exploratory piece will be more about the readers doing the analysis themselves, putting the effort in to discover things that strike them as being significant or interesting.

In the following image, we see a **scatterplot matrix** visualization: a method used to reveal correlations across a multivariate dataset, enabling the eye to efficiently scan the entire matrix to quickly identify variable pairings with strong or weak relationships. This is a perfect example of an exploratory visualization design:

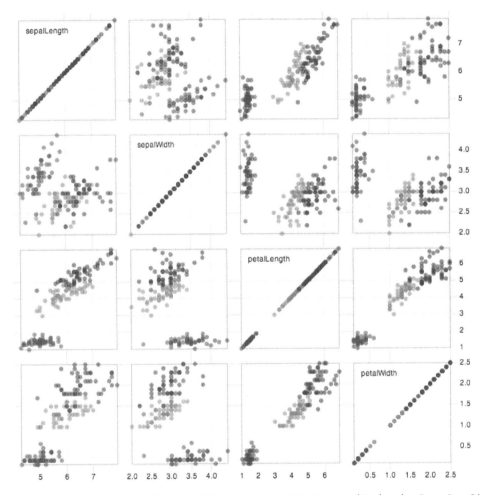

Image from "Scatterplot Matrix" (http://mbostock.github.com/d3/ex/splom.html), created by Mike Bostock.

Exploratory visualizations are not limited to being interactive. Visual analysis can be facilitated through static portrayals of data. The previous example is actually interactive but a static version would still offer a discovery of the relationships and patterns of the dataset.

That said, it is fair to say that in order to create a truly exploratory experience, interactivity does introduce the potential for so much extra functionality to help immerse the user into a dynamic, problem-solving challenge.

Features such as filtering, sorting, brushing (selecting or isolating certain data values), variable adjustment, and view modification are just some of the important ways you can help a user investigate data. We'll discuss more about interactivity in *Chapter 4, Preparing and Familiarizing with Data*.

It is also worth highlighting that while explanatory visualization is primarily created for others, exploratory data and the process of visual analysis can be as much for your own discovery purpose as it is for others. It is clearly a particularly relevant function for scientists, for example, to find patterns and unearth key findings in research work before the publication of results (which would then require the use of explanatory-based visual evidence).

When the function is to exhibit data

The final classification of intended function is in some respects a controversial one, because including the exhibiting of data as an intended function of visualization will not be consistent with many people's definition of data visualization.

We're not talking controversy on the level of a political scandal. Let's put it into some perspective: within the context of this field, this is a big deal and the cause of so much debate.

As with any attempt at classification, there is a spectrum of variety within and so clear boundaries are difficult to establish and very much open to personal interpretation.

We are talking here about designs that use data as the raw material, but where the intention is perhaps somewhat removed from a pure desire to inform. Rather, the objective is closer to a form of exhibition or self-expression through data representation. This genre of work embodies the term "data art".

Data art is characterized by a lack of structured narrative and absence of any visual analysis capability. Instead, the motivation is much more about creating an artifact, an aesthetic representation or perhaps a technical/technique demonstration. At the extreme end, a design may be more guided by the idea of fun or playfulness or maybe the creation of ornamentation.

This particular strand of data visualization is contentious simply because it challenges those seeking to identify the boundaries of this field and its proximity to other disciplines such as graphic design, generative design, or creative art.

In the following example, we see an example of data art (as defined by the creator himself) that visualizes all the adjectives used in Cormac McCarthy's book *The Road*. The adjectives are arranged radially in alphabetical order and each line represents a timeline of the book, beginning at the perimeter:

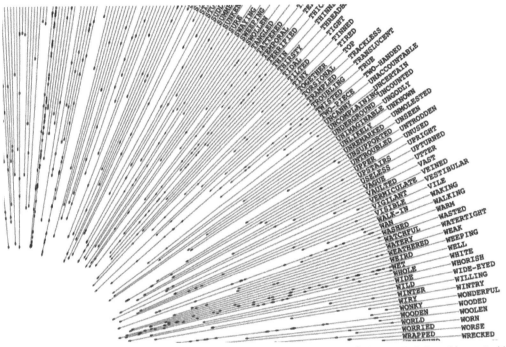

Image from "Adjectives of The Road" (http://distantshape.com/the_road.html), created by Kemper Smith.

The effect is an interesting artifact to look at and its construction is representative of an impressive technical or algorithmic solution, but its primary intent is not to easily allow us to learn about the language in the book. I would personally position this right on the boundary between an exploratory visualization (static, but allowing readers to look up combinations of adjectives and see patterns) and data art.

We characterized explanatory visualization as a single visual experience and exploratory as a numerous but finite set of experiences. By contrast, the range of reactions to exhibition-type designs has a more open and unlimited potential. It aims for and triggers more of an aesthetic reaction, which means our emotional connection and interpretation can vary significantly from one person to the next.

Establishing intent – the visualization's tone

Earlier, we looked at a collection of different phrases that articulate realistic intentions for creating a data visualization. We've just proposed three classifications for how you might organize these terms according to their function.

There will sometimes be an overlap between these otherwise distinct categories, but you should be able to determine relatively clearly where your work intends to fit on the scale between explanation, exploration, and exhibition.

Setting the function is just one part of the "intent" equation. The clarity of your potential design pathway will be much more apparent as we now consider the second dimension of intent—tone.

Establishing a suitable tone goes beyond function and more towards the style of the design experience. It concerns the type of stimulus or desired emotional response that you are trying to create. It is therefore important for you, as the designer, to be able to reason what sort of design will achieve that tone.

With this type of judgment to make, you will inevitably find yourself juggling creative and scientific perspectives. This dynamic poses a significant challenge for any data visualization designer to reason and resolve. Indeed, the design objectives outlined at the end of the previous chapter highlight the intricate issues you have to handle.

I've alluded to the debates and arguments around data art, well this subject inspires the most prominent dividing lines and debates that exist within the field revealing the occasionally awkward and misunderstood relationship between art and science, the pillars of this field.

While it is possibly a crude generalization, the standpoint of the science side is characterized as being concerned with preserving the efficiency and accuracy of judgments derived from a visualization. Variations in data representation that steer away from this goal are believed to reduce the quality and effectiveness of a visualization.

On the other side, the artistic section of the community can be viewed as being concerned with experimentation, finding creative expressions of data, and new aesthetic connections with an audience. Practitioners in this cohort, typically, will have arrived in the field from a strong, design-led or computer science background.

For the rest of us, somewhere in the middle, we find ourselves either holding hands with both sides, following the direction of the wind from one day to the next, or simply sat on a fence without any strong favor towards either side.

So how can we rationalize the role of these two very different, opposing perspectives and beliefs? In my view, the latter enhance the field by demonstrating what can be achieved through the aesthetic and technological creativity. The former help us understand what we should do through the pursuit of evidence and observation of rules around human cognition and visual perception. We need visualizations that look appealing and we need visualizations that perform well.

However, sometimes there has to be mutual recognition that for different scenarios there might be good reason for leaning more towards one direction than the other.

Let's look at the language of two potential motives behind creating a data visualization:

- "We need a chart to help monitor..."
- "We need to present this in a way that persuades people..."

Here, we have two situations both aiming to better inform a reader or a user, but the intended effect or outcome from the experience will be different.

The reaction of a user reading, for example, a dashboard full of bar charts and line charts to help monitor monthly performance will be quite analytical and pragmatic in style. It is unlikely to involve or stir much emotion (unless things are suddenly and unexpectedly plummeting). The style of the visualization design will be consistent with the intended nature of this particular type of engagement, probably quite sober and with an emphasis on the precision of perception.

Compare that with the intended impact of a presentation that depicts how many lives could be saved if a charity was able to achieve a certain level of fundraising. The setting and intent will be more about persuasion making it emotionally charged. It will need to attempt to create an experience that is much more personal and more impactive.

Here, we see potentially two ends of a spectrum for judging the right tone. Yet they fundamentally share the overall motivation of wishing to inform people about a subject through the visual representation of data.

One scenario would achieve this in a relatively pragmatic style, influenced by a desire to optimize the efficiency and accuracy of interpretation. The other would be judged as effective if it evoked a suitably positive emotional response to the data story.

We can therefore describe tone as being a continuum from a pragmatic or analytical portrayal through to a more emotive or abstract concept.

Pragmatic and analytical

The following is a quotation from Jock Mackinlay (`http://hci.stanford.edu/courses/cs448b/f10/lectures/CS448B-20100923-DataAndImageModels.pdf`):

"A visualization is more effective than another visualization if the information conveyed by one visualization is more readily perceived than the information in the other."

This quote perfectly captures the priority and intent behind pragmatic or analytical visualizations. Some might term them simple or boring but that is short-sighted and lacking in appreciation for the setting in which these types of data portrayal is vital.

Designs that fit this classification will often involve data being represented through the use of bar charts, line charts and dot plots, for example. Stylistically, they will be characterized by a rather clinical look-and-feel that is consistent with the next sample image, taken from a project analyzing Olympic results over the years:

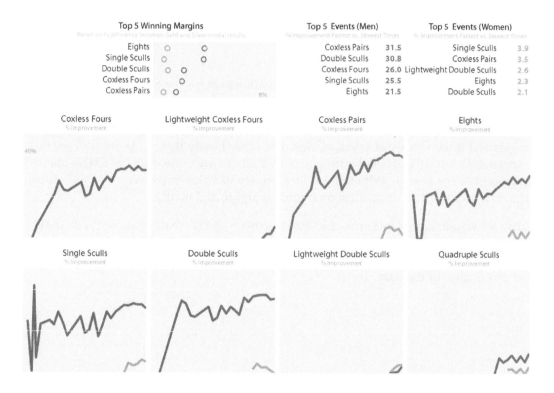

Creating a visualization with a pragmatic tone is about recognizing a need for a design that delivers fast, efficient and precise portrayals of data. Typically, you will have a captive audience, a readership who want to or need to interact and learn from the data. This could be a corporate environment, where people need to simply learn about recent performance of operational activity or undertake visual analysis to discover potentially revealing patterns.

In these cases, there is no value placed on attempting to draw attention to the visualization, or trying to encourage somebody to read or interact with a graphic by employing aesthetic novelty. Furthermore, it's not about trying to inject any emotional or metaphorical connection with the data stories presented.

The purity and impact of its function—the satisfaction that comes from an efficient intake of understanding—fulfils the aesthetics of the charting methods deployed. Therein lies the elegance of pragmatic work.

Emotive and abstract

The following is a quotation from Chris Jordan, *TED2008* (`http://www.ted.com/talks/chris_jordan_pictures_some_shocking_stats.html`)

> *"I have a fear that we aren't feeling enough, we aren't able to digest these huge numbers."*

At the other end of the spectrum are visualizations where the tonal intent is much more emotive and/or abstract in form.

Sometimes you just want to and need to move beyond bars, straight lines, and right angles and more towards curves, circles, and other bendy things. As we will see in *Chapter 4, Preparing and Familiarizing with Data*, there are consequences to this choice, in respect of the known reduction in the accuracy of value perception this will cause. That is a sacrifice you as the designer need to juggle and justify.

Abstract visualization, in terms of its tone, is more about creating an aesthetic that portrays a general story or sense of pattern. You might not be able to pick out every data point or category, but there is enough visual information to give you a feel for the physicality of the data.

This next image is taken from a project to visualize the global airline transportation network consisting of all commercial flights worldwide. The routes highlighted are those flights in and out of Toronto Pearson airport. The project was designed to assess the threat of infectious diseases.

Image from "Toronto Flight Lines" (http://www.biodiaspora.com/) created by Bio.Diaspora 2012

The design does not intend to offer an analytical summary of air travel statistics. Instead it creates a more immersive experience in to the data, offering a visual interface to establish a greater sense of how interconnected the world is through air travel. It causes us to imagine just how easy it could be for diseases to spread across the globe in a short period of time.

For more emotive visualizations, you might be seeking to generate a different type of emotional connection with the design. This connection comes both at the start of the engagement—creating attraction and appeal—as well as after the engagement—the outcome.

In contrast to pragmatic works, as we described in the introduction, here we might be seeking to achieve impact, to emphasize issues, and perhaps to engender shock. We might also seek to generate a certain amount of visual attention through novelty and innovation in a way that more pragmatic approaches would not be able to achieve.

In the following image, we see a section taken from a newspaper infographic that depicted Iraq's bloody toll. While the chart method is nothing more complex than an upside down bar chart, the tone is very impactive and metaphorically emphatic, creating a strong emotional impact to the story it portrays:

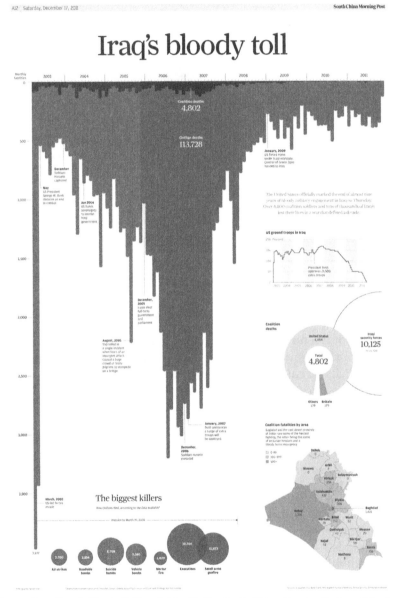

Image from "Iraq's Bloody Toll", published in the South China Morning Post on Saturday, December 13, 2011 (http://graphics-info.blogspot.hk/2012/09/malofiej-20-look-at-our-participation.html), created by Simon Scarr.

Of course, it is important not to stretch the functional and tonal responsibility and capability of visualization too far. This is where any hype and disproportionate expectation about the potential impact of data visualization can be misplaced.

Data visualization is a means to an end, not an end in itself. It's merely a bridge connecting the messenger to the receiver and its limitations are framed by our own inherent irrationalities, prejudices, assumptions, and irrational tastes. All these factors can undermine the consistency and reliability of any predicted reaction to a given visualization, but that is something we can't realistically influence.

All we can do is form a best judgment about where on the continuum of design style, from a pragmatic experience through to an emotive one, the purpose of our data visualization will be most suitably defined. The ultimate responsibility for what happens beyond the visualization engagement sits with the reader or the user.

Key factors surrounding a visualization project

The following is a quotation from Edward Tufte's book, *The Visual Display of Quantitative Information*:

> *"Most principles of design should be greeted with some skepticism... we may come to see only through the lenses of word authority rather than with our own eyes."*

While establishing the purpose of the visualization project sets the desired tone of the design and its function, there are inevitably many other factors that will have a significant influence on the shape and direction of our visualization design.

It is especially important to identify and recognize the impact of the contextual conditions, within and around your project that will affect what you can and can't achieve and how you might achieve it.

This list of factors may seem quite obvious and fairly rudimentary, but if we wish to eradicate the likelihood of misjudgments or misunderstandings, and maximize the efficiency and effectiveness of the process, we need to nail them early on.

There is simply no point waiting until it is too late to consider these, because by then you will have already followed a certain path and spent valuable time and resources on your work.

Here are some of the most important factors to consider and to evaluate their potential impact:

- **The aim**: As we have seen already, there are different origins and triggers for a project. We mentioned the self-initiated ones as being almost free of external constraint and essentially framed by our own capabilities and intentions. The important thing worth reinforcing here is the need to take responsibility when a project involves a brief, commissioned by a client or a colleague. You must demonstrate excellent communication skills to ensure you seek and gather as much of an understanding as possible of what it is they are aiming to achieve. Sometimes, you might be provided with a very open brief because a client may not even know what it is they are seeking. In these situations, your responsibility needs to extend to assist them in the scoping and requirements of the work. On other occasions you will be asked to create something that goes against your general practice (for example, the subject matter or requested style) or it might even be simply impossible to deliver (perhaps due to the desired design or available resources). Here again your communication skills are going to be required to manage the expectations. It is easy to be shy and delay asking vital questions but this will only cause you pain later.

- **Time pressures**: Common to just about every commissioned design project will be the pressure of time and deadlines. Most projects have clear timescales, from in-day quick turnaround pieces to longer-term grand projects. The challenge of maintaining objective creativity in the face of diminishing time is something that will severely test designers of all experiences. Whatever your situation, you have to use your time effectively and that's where value will come from following the tactics in this methodology. Plan your work and create a balanced layout of the things you need to accomplish, so that you avoid disproportionately spending time on tasks that are less important than others. Often you will find yourself undertaking a visualization project in parallel with many other commitments. Not only will your capacity be limited, the momentum and duration of your focus will be impacted. This is where project management skills come to the fore as well as a realistic appreciation of what you should and shouldn't commit to undertaking. It also highlights the importance of keeping notes so that you can move seamlessly between projects and not lose track of your thoughts, ideas, or progress.

- **Costs**: The issue of financial resource will unquestionably emerge, especially for large-scale projects. Costs will significantly influence the time you are able to commit to a project, the scope for bringing in additional collaborators, and the range of tools or technical resources you might be able to utilize. Once again, the planning and preparation stages will be invaluable to surface all potential issues around financial matters.

- **Client pressures**: Aside from time pressures, you need to anticipate and reduce the impact of potential unexpected pressures and interruptions coming from your client or colleagues. This might be changes in requirements, new demands, interference in the design solution, and generally annoying things that get in the way of your progress. A further manifestation of the pressure that can come from clients is the insistence on observing organizational visual or brand identities, layout rules, editorial guidelines, and technical frameworks. All of these will shape the scope of your design choices. You have to be prepared for and capable of managing this relationship, and the mutual expectations, effectively so always be open with your client, keep them regularly updated with progress and, where applicable, involve them in the key decision moments throughout the process.

- **Format**: From a design perspective this will be a significant influencing aspect. Are you creating a static or an interactive design? Maybe it's a multifaceted project and you are looking to create both. If it is an interactive design, what platform do you need to achieve compatibility with? Will it be for the Web, a tablet, and/or smartphones? If it is a static design, will it be a small graphic in a publication, a full-page spread, or a large poster display? Maybe it will be a video animation or an ambient display out in the wild, or a large touch screen installation in a museum. This is a vital consideration that needs to be cleared up at the earliest possible stage. Another factor to take into account will be the likely frequency of the project — is it a one-off piece or will it be something that needs to be replicable and/or scalable? That could hugely affect what you can or can't deliver.

- **Technical capabilities**: Aside from your own technical capabilities, what are the technical resources to which you have access? For example, are you limited to free tools or can you access more premium software? Do you have the most appropriate technical infrastructure, such as server speed and capacity if it is an online project? Depending on your format choices, what frameworks are you going to deploy, what browsers do you need to have it working on, what backend database technologies are you going to require? This is a wide-ranging and very technical set of decisions that will likely require a specialist technician to determine.

The "eight hats" of data visualization design

The final scoping issue to consider at this stage of your visualization design project is an assessment of your personal capabilities and those of any collaborators that you involve in the work. What skills and knowledge do you collectively possess or lack? This is a big issue for many, so we need to spend the remaining pages of this chapter looking at it closely.

The demands on a visualization designer in terms of capability are many, reflecting the truly multidisciplinary nature of the subject. The convergence of different ingredients introduces a wonderful richness and variety of issues to be concerned with, but it can equally present quite a challenge for people looking to master the subject.

For many, the prospect of trying to acquire the necessary array of knowledge and skill across the entire range of capabilities is something that can be intimidating or at least exist as a perceptual barrier. There is a sense that to be successful you need to be some sort of superhero.

Taking an analytical look at the range of required capabilities reveals a role and need for many types of people, which can of course be fulfilled by a number of people or just one.

These are proposed as the "eight hats of data visualization design". Influenced by the concept of Edward de Bono's six thinking hats, which related to the different thinking perspectives we should try to occupy when tackling complex problems, this is an attempt to organize the different attributes required to accomplish success in visualization.

It should help you recognize where you fit it in to the spectrum of duties and responsibilities, helping you identify your strengths and your weaknesses accordingly. You may then choose to address these weaknesses personally or plug the gaps with support from others.

The initiator

The **initiator** is the leader, the person who is seeking a solution to the task as per the brief or self-initiated curiosity. The hat is that of an explorer; they want to explore data and different design avenues to find answers to problems or evidence to serve their researcher mindset. The initiator will be responsible for much of the considerations covered in this chapter. They will establish the functional and tonal direction of the project, as well as identify and profile the target audience. The initiator will also define other parameters such as the intended format/platform of the solution and some of the key technological issues.

The data scientist

The **data scientist** is characterized as the data miner, wearing the miner's hat. They are responsible for sourcing, acquiring, handling, and preparing the data. This means demonstrating the technical skills to work with data sets large and small and of many different types. Once acquired, the data scientist is responsible for examining and preparing the data. In this proposed skill set model, it is the data scientist who will hold the key statistical and mathematical knowledge and they will apply this to undertake exploratory visual analysis to learn about the patterns, relationships, and descriptive properties of the data.

The journalist

The **journalist** is the storyteller, the person who establishes the narrative approach to the visualization's problem context. Working with the data scientist and the initiator, they are able to establish the key stories and angles with which to slice the analysis. They work on formulating the data questions that help keep the project's focus on its intended editorial path. Building on the initiator's initial sparks of ideas, the journalist will develop a deeper researcher mindset to really explore the analytical opportunities.

The computer scientist

The **computer scientist** is the executor, the person who brings the project alive. With their critical technical capability they are ultimately the ones who will construct the solution. They will also bolster the data scientist with their technical know-how to most effectively and efficiently handle the data gathering, manipulation, and pre-production visualization activities. The breadth of software and programming literacy will have a great bearing on the potential direction and sophistication of the data visualization solution, whether this is created within a tool or through programming.

The designer

The **designer** is the creative, the one, who, in harmony with the computer scientist, will deliver the solution. They have the eye for visual detail, a flair for innovation and style and are fully appreciative of the potential possibilities that exist. However, they also have the necessary discipline to follow the message established by the initiator and taken on by the journalist. They respect the capabilities of the computer scientist in terms of what solutions could be feasible, but themselves have the helicopter-like vision to rationalize and reason what things will work and will not work, and why.

Their key responsibility is also to be capable of ensuring the harmony of the solution between its form and its function, ensuring it is aesthetically appealing to draw in the reader while fundamentally delivering the intended, communicated message.

The cognitive scientist

The **cognitive scientist** is the thinker in terms of appreciating the science behind the effectiveness of the technical and designed solutions. They have the visual perception knowledge about how the eye and the brain work most effectively and efficiently. They also have deep knowledge about concepts such as the Gestalt Laws, communication theories, color theories, and human-computer interaction principles. Additionally, they are able to inform the design process in relation to the complexities of how the mind works in terms of memory, attention, decision-making, and behavioral change.

The communicator

The **communicator** is, naturally, concerned with the communication side of the project. With their hard hat on, they act as the negotiator and presenter, operating at the client-customer-designer gateway, helping to inform all those who are involved on progress, requirements, problems, and solutions. The communicator needs to be close to all stages of the process, understanding requirements, appreciating restrictions, recognizing possibilities, and then ultimately launching, publicizing, and showcasing the final work. An ability to articulate and explain matters to different types of people, technical and non-technical, and be capable of managing expectations and relationships is vital.

The project manager

This final role is essentially that of the manager or coordinator, the person who does much to pick up many of the unpopular duties to help bring the whole project together. They manage the project's process and its progress, ensuring it is cohesive, on time, and on message. They understand the brief and identify/manage all the key factors surrounding the project. Ultimately, this role is required to ensure things get finished, so they need to have an eye for detail, the commitment and patience to check everything and they should also be concerned with integrity matters around visualization ethics.

Summary

In this chapter, we have started our journey through the data visualization methodology. The emphasis has been on the importance of planning, preparation, and scoping our project, before we embark on any design work. Without this early work we could undermine the effectiveness and efficiency of our eventual design process: something any designer can ill afford to allow.

We have seen how data visualization is a means of facilitating the discovery of patterns and relationships that exist within data. These are insights that would otherwise be practically impossible to draw from data in its raw state.

The importance of establishing the purpose of our visualization project was the key part of this first stage. Specifically, we highlighted the distinction between functional intent and tonal intent.

Within these characteristics we described the difference between visualizations that are functionally seeking to explain, explore, or exhibit data. Furthermore, we saw the significance in potential design differences between visualization styles that serve a pragmatic tone and those that are more emotive or abstract.

As we will appreciate throughout the remaining chapters, developing the clarity of our purpose at this early stage is paramount to the success of our visualization design process. The choices we make fundamentally influence our design choices and the potential experience of our target audience.

We explored some of the key factors that can have a strong influence on the shape and scope of our visualization project. Whether it is the technical matters, the issue of format, financial resources, or timescales, each factor mentioned can have a huge impact on your creative path and scope.

Finally, we looked in depth at the range of personal capabilities required to successfully deliver a visualization design and drew attention to how you might need to personally address any gaps through development or collaboration.

In the next chapter we'll look at two further important stages of planning and preparation: identifying your intended narrative and getting intimate with your data.

3
Demonstrating Editorial Focus and Learning About Your Data

In the previous chapter, we introduced the data visualization methodology, starting off with a look at two important preparatory activities: establishing the project's purpose and identifying the influencing factors surrounding the project.

It is worth acknowledging that the intention of our visualization may evolve, particularly as we journey through the upcoming design phases and as new influencing factors emerge. Any decisions we make across this process can be revisited and refined but the greater clarity we achieve now will ultimately help minimize wasted efforts and lead to a more efficient process.

In this chapter, we move on to the next phase of the methodology where we look at the intertwining issues of the data we're working with and the stories we aim to extract and present. This activity provides a bridge between project inception and design conception and involves the following tasks:

- We will look to develop and refine our editorial focus around the key communication dimensions of our visualization problem: What is the story we are trying to tell? What is the key narrative we are looking to portray? What questions do we wish readers to be able to answer through the visualization?

- One of the biggest challenges, and usually the most time-consuming, is the acquiring and preparing of the data, ensuring it is fit for purpose, and in good shape in advance of the design stage. We'll explore the mechanics of working through this often hidden activity.

- Finally, we'll see an example of how we can use visual analysis techniques to combine the task of familiarizing with our data and discovering key insights. We will show how learning about the physical properties of data helps you develop your editorial focus, specify your data questions, and influence the potential design choices we make later on.

The importance of editorial focus

The following is a quote from Edward Tufte (`http://adage.com/article/` `adagestat/edward-tufte-adagestat-q-a/230884/`):

> *"Good content reasoners and presenters are rare, designers are not."*

In *Chapter 2, Setting the Purpose and Identifying Key Factors*, we looked at some of the considerations involved in identifying the purpose behind your visualization project; these are as follows:

- What is the reason for its existence?
- For whom are we creating it and how well defined are the requirements?
- What function is it seeking to fulfill?
- What is the likely tone of the design we're intending to portray?

Over the course of the full design process, it is possible that these initial definitions may need to be modified. As we learn more deeply about the relationship between what we want to do, what we can do and, importantly, what we should do, our creative proposition may be molded into a slightly shape.

That's fine and is to be expected. However, the earlier we can make firm judgments on our creative direction the better. This gives us a solid starting point and helps inform the important decisions we need to make about what it is we are trying to say with the visualization we are developing.

The matter of *how* this is said will be covered in the design stage but, ahead of that work, we first need to determine *what* are the specific messages we are looking to communicate to our audience.

Some of the most influential and esteemed visualization and infographic design work, perhaps unsurprisingly, comes from newspaper and magazine organizations.

The New York Times would probably be at the top of many peoples' list of the most celebrated graphics work, but there are so many other examples of great innovation and excellence from across the industry and right around the world, including The Guardian (UK), National Geographic (US), the Washington Post (US), the Boston Globe (US), La Informacion (Spain), and Época (Brazil), to name but a few.

A key reason behind the success of the work produced by these departments is the demonstration of what Edward Tufte describes at the beginning of this section—editorial focus.

Regardless of the size and inherent complexity of the data challenge you are working on, this is one of the most important capabilities you will need to develop in order to succeed in data visualization and is something that can singularly influence the success, or otherwise, of a design.

An editorial approach to visualization design requires us to take responsibility to filter out the noise from the signals, identifying the most valuable, most striking, or most relevant dimensions of the subject matter in question.

To do this we need to weigh-up the potential appetite of the intended audience—what it is we think they will want to know or will find interesting—and the opportunities that exist within the data—what data stories can you find and might you portray.

Determining what an audience needs is not always straightforward, particularly when you might have a broad range of different types and background of readers engaging and interpreting your work. Nevertheless, you should still have a sufficiently sympathetic view of how your target demographic will most positively and constructively relate to different slices of analysis of your subject matter.

For projects triggered by a client or colleague, there may be specific analytical dimensions that are already established and you have been asked to present and communicate them. The scope for veering away from this existing focus may not exist.

Otherwise, irrespective of whether you are tasked with the work or just pursuing a self-initiated curiosity, in most cases, you will have a certain degree of liberty to undertake the dual role of analyst and storyteller influencing the selection of what you will portray.

The execution of a design is clearly hugely significant to the success of a project, but without the foundation clarity and justification for the message you are trying to communicate, your resulting visualization will fundamentally lack focus.

Rather than just throwing everything available at a reader, good visualization involves showing a degree of editorial care—just because you have some data, doesn't mean to say you have to use it all. Be selective.

This attitude is necessary for all types of visualization projects. You might think the idea of telling stories is only relevant for explanatory pieces. That's not the case. With exploratory designs you still need to demonstrate this editorial focus. The difference is that with these projects you are not so much telling stories rather you are making them accessible and discoverable. You still need to frame the subject matter and define the important dimensions of analysis that will be made available for manipulation and interrogation. You still need that level of care for the audience's interpretive experience.

As we'll see later, some of the most effective data visualization designs manage to create a combination of these functional characteristics, offering a sweet spot of engaging exploratory features framed within defined story dimensions.

Conversely, if you take a look at a gallery of visualization work and find examples that you believe are ineffective, they will likely exhibit a weak narrative, an absence of stories, and a lack of genuine care for the interpretive needs of the audience. This is a really influential dimension of visualization design.

Preparing and familiarizing yourself with your data

The following is a quote from Simon Rogers, *The Guardian, Facts Are Sacred: The Power of Data*:

> *"80% perspiration, 10% great idea, 10% output."*

Before we get too far down the line of developing and defining our intended stories and analytical slices, we need to roll up our sleeves and get our hands dirty with the task of accessing and preparing our data.

Whether you get the data first or shape your desired story dimensions first is mainly going to be influenced by the context of your project. It is a somewhat "chicken and egg" situation—which comes first, the data or the focus? You need some focus to determine what data you need, but you don't know what potential insights exist in the data until you have it.

It is best to accept that there will be a certain amount of iteration as you alternate between the mindset of a data scientist and a journalist progressing both issues simultaneously.

Data is our raw material, the principle ingredient in our creative recipe. Irrespective of what we intend or hope to show through our visualization design, the data will ultimately do the talking.

If we don't have the data we want, or the data we do have doesn't tell us what we hoped it would, or the findings we unearth aren't as interesting as we wish them to be there is nothing we can (legitimately) do about it. That is an important factor to remember. No amount of 3D-snazzy-cool-fancy-design dust sprinkled on to a project can change that.

An incomplete, error strewn or just plain dull dataset will simply contaminate your visualization with the same properties. So, the primary duty for us now is to avoid this happening, remove all guessing and hoping, and just get on with the task of acquiring our data and immerse ourselves into it to learn about its condition, its characteristics, and the potential stories it contains.

To achieve this, we must go through the often painful mechanics of data familiarization and preparation; as follows:

Acquisition: First, you need to get hold of your data. As we have discussed, this might already be provided to you from those commissioning the work. You might have independently formed a sense of the specific subject dimensions on which you require data. Alternatively, it may be that you have yet to focus beyond a broad subject level. It all really depends on how well-defined your requirements or intentions already are.

The places where you might acquire your data and the methods to accomplish it will be something you will know best. It could come from origins such as these:

- Obtained from a colleague, client, or other third-party entity
- A download taken from an organizational system
- Manually gathered and recorded
- Extracted from a web-based API
- Scraped from a website
- Extracted from a PDF file (you have my sympathies)

Here, we see an image taken from a visualization project that was created to demonstrate the social expansion of the US using the story of the spread of post offices. In this screenshot of the final piece, we see a representation of the 11,000+ post office locations recorded across the country between 1700 and 1900:

Image from "Posted: Visualizing US Expansion Through Post Offices" (http://blog.dwtkns.com/2011/posted/), created by Derek Watkins

The entire data for this project was scraped from the US Postal Service website. After cross-referencing the dataset with a gazetteer to establish accurate geo-locations, almost 1,500 records (12 percent) had to be discarded, as they weren't readily "mappable".

This just shows the great amount of effort and pain that often goes in to sourcing and preparing your data. No matter from where you are accessing your data, you will often have to work hard to get it into the shape and form that you need it. Therefore, you need to ensure you have factored in as much time as possible for this vital stage of the process.

Examination: Once we've got the data, a thorough examination will determine your level of confidence in the suitability of what you have acquired. This involves assessing the completeness and fitness of the data to potentially serve your needs. There are many tools out there that can help you work through this stage efficiently. Depending on the size and complexity of your data, and obviously your own capabilities, software like Excel, Tableau, or Google Refine (among plenty of others), will enable you to quickly scan, filter, sort, and search through your dataset to establish its state of quality. As you go through this process, you should be examining the following potential issues:

- **Completeness**: Is it all there or do you need more? Is the size and shape consistent with your expectations? Does it have all the categories you were expecting? Does it cover the time period you wanted? Are all the fields or variables included? Does it contain the expected number of records?

- **Quality**: Are there noticeable errors? Are there any unexplained classifications or coding? Any formatting issues such as unusual dates, ASCII characters? Are there any incomplete or missing items? Any duplicates? Does the accuracy of the data appear fine? Are there any unusual values or obvious outliers?

Data types: Understanding the properties of our raw material is such an important task. We will do some visual exploring later to learn about the physical patterns and relationships but, for now, we need to understand the fundamental structure of our data in terms of the variables types. This will become important when we move into the design discussion in *Chapter 4, Preparing and Familiarizing With Data*. The following table outlines the discrete types of data with associated examples:

Types	Examples
Categorical nominal	Countries, gender, text
Categorical ordinal	Olympic medals, "Likert" scale
Quantitative (interval-scale)	Dates, temperature
Quantitative (ratio-scale)	Prices, age, distance

As well as capturing the types of data we have, it is a useful exercise to also make a note of the range of values or at least a sample of the data held against each field. For illustration, this might be from a dataset about the Olympics:

Data	Types	Range
Event	Quantitative (interval-scale)	27 different years (1896–2012)
Medal	Categorical ordinal	Gold, silver, bronze
Athlete	Categorical nominal	1500+ different athlete names

Data	Types	Range
Result	Quantitative (ratio-scale)	Race results (9.59s > 4:02:59)
Country	Categorical nominal	96 different country names

Transforming for quality: This task is naturally about tidying and cleaning your data in response to the examination stage above. We are looking to resolve any of the errors we discovered in order to transform the condition of the data we're going to be working with for our design. Plugging the gaps caused by missing data, removing duplicates, cleaning up erroneous values, and handling uncommon characters are just some of the treatments we may be required to apply.

Transforming for analysis: In contrast to transforming for quality, we move away from cleaning data and focus more on preparing and refining it in anticipation of its intended use for analysis and presentation. Here, we consider actions such as:

- Parsing (split up) any variables, such as extracting *year* from a date value
- Merging variables to form new ones, such as creating a whole name out of *title*, *forename*, and *surname*
- Converting qualitative data/free-text into coded values or keywords
- Deriving new values out of others, such as *gender* from *title* or a sentiment out of some qualitative data
- Creating calculations for use in analysis, such as percentage proportions
- Removing redundant data for which you have no planned use (be careful though!)

Another important consideration is to determine what level of resolution you might wish to, or indeed need to, present your data. The decision you take about this may require you to aggregate or disaggregate your data to achieve get the right level of detail.

Design agency Periscopic were faced with some intricate resolution decisions in their preparatory work for this near real-time visualization developed about the Yahoo! Mail network. The objective was to show the huge volumes of e-mails being sent and processed around the world at any given point, and the efforts Yahoo! is taking to help reduce and intercept spam e-mails. This is shown in the following screenshot:

THE **Yahoo!** MAIL NETWORK IS DELIVERING **57,520** EMAILS PER SECOND WORLDWIDE.

Image from "Visualizing Yahoo! Mail" (`http://www.periscopic.com/#/work/yahoo-mail/`),
created by Periscopic

With approximately 5.6 billion e-mails (and a further 20.5 billion spam) sent every day, the sheer amount of data potentially being fed into this project clearly posed a challenge in terms of what level of detail they could reasonably show.

This was not just a matter of how they could handle the velocity and volume of new data on the technical side but also what was the appropriate resolution with which to tell this story

They decided on the following strategy:

- The headline statistics shown in the titles and presented across a range of supplementary graphics across the project would be representative of the full data quantities.

- For the geo-spatial view, a carefully designed algorithm was applied to extract a representative sample of data. This would be more than adequate to capture the nuances of the activity seen with the full dataset and would avoid the technical impracticalities involved in attempting to show 100 percent of the data.

- The geographical data was clustered to a city or regional aggregate, represented by the circle positions and sizes, to help draw out the key signals and patterns.

This is a perfect demonstration of how important it is to handle data resolution issues as early as possible so we know what treatment to apply to our data.

When you are faced with similar decisions, albeit perhaps rarely on the same scale, you will typically have these options available to you:

- **Full resolution**: Plotting all data available as individual data marks.
- **Filtered resolution**: Exclude records based on a certain criteria.
- **Aggregate resolution**: "Roll-up" the data by, for instance, month, year, or specific category.
- **Sample resolution**: Apply certain mathematical selection rules to extract a fraction of your potential data. This is a particularly useful tactic during a design stage if you have very large amounts of data and want to quickly develop mock-ups or test out ideas.
- **Headline resolution**: Just showing the overall statistical totals.

Consolidating: When you originally access your data, you will likely believe, or hope that you have everything you need. However, it may be that after the examination and preparation work, you identify certain gaps in your subject matter.

Additional layers of data may be required to be combined ("mashed-up") with our existing dataset, applied to perform additional calculations, or just to sit alongside this initial resource to help contextualize and enhance the scope of our communication. Always spend a bit of time considering if there is anything else you anticipate needing to supplement your data to help frame the subject or tell the stories you want to communicate.

Seasoned designers will confirm that acquiring, handling, and preparing your data is often the most time-consuming and intensive activity involved in any visualization project.

It is the hidden battle. As Simon Rogers quantifies at the start of this section, if you imagine a visualization design project as an iceberg, the final design would be the bit we see sticking out of the water and the ugly data preparation work would be the vast amount hidden beneath the surface.

There is a good chance that you will have expended most perspiration in the many thankless, uncelebrated duties you have to undertake in this part of the process. However, just know that the value of your efforts and the associated rewards will emerge in due course, so try not to lose enthusiasm or patience.

Refining your editorial focus

The following is a quote from Amanda Cox, New York Times (http://seekingalpha.com/article/66269-an-amazing-graphic-on-box-office-receipts):

> *"Different forms do better jobs at answering different questions."*

Now that we have prepared our data, we revisit the matter of editorial focus.

To avoid being prematurely tempted into diving into the construction of a visualization design, we first need to do more work to fine-tune our analysis of what are the important messages.

In the first section, we explained the importance of taking responsibility to make sense of data, to find stories and tell stories to your intended audience. This demonstrates a level of care. It shows that you are not just going through the motions of communicating; you are taking it seriously, seeking to help your audience unlock insights from the subject matter.

The journalistic capability for unearthing the most relevant stories from data is a talent that any designer should aspire to develop.

In the example shown in the following screenshot, we see a recent visualization project that was developed to enlighten people about the matter of education around the world, presenting some striking facts and figures:

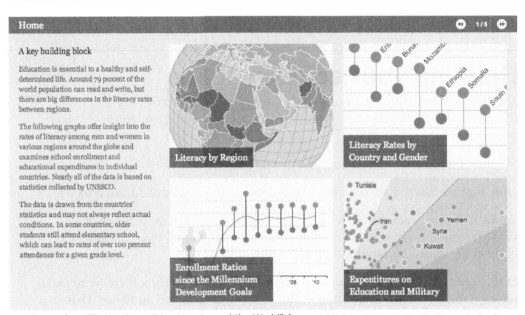

Image from "In Numbers: Education Around the World" (http://visualdata.dw.de/specials/bildung/en/index.html), created by Gregor Aisch for Deutsche Welle.

As you might imagine there will be myriad ways of telling data stories about global education matters. In such contexts, a designer is faced with the challenge of rationalizing so many different potential dimensions.

The strength of this particular project comes from the scoping and definition of the chosen narrative and slices of analysis. Rather than bombarding the reader with endless pages of facts and figures, or offering seemingly infinite combinations of interactive variable selections, the subject is framed for us around a small number of interesting angles about education: literacy by region, literacy rates by country/gender, enrollment ratios, and expenditure on education versus military.

As we then navigate through each story panel we are presented with a series of explanatory visualizations. They don't just show data, they present and explain it.

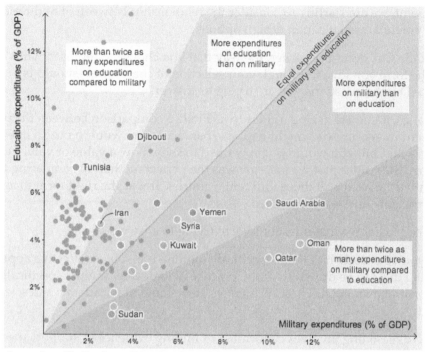

Image from "In Numbers: Education Around the World" (`http://visualdata.dw.de/specials/bildung/en/index.html`), created by Gregor Aisch for Deutsche Welle

In this example, we see a scatter plot of education spend versus military spend for all countries. But it is more than just a plot. The designer takes responsibility for telling the story, providing effective written (labeling and captions), and visual annotation (reference lines and background shading) to help maximize the potential insights. The inclusion of filtering features to highlight particular countries and regions introduces an exploratory dimension to enable the discovery of further layers of understanding.

This is a strong demonstration of editorial focus and storytelling with data—four key stories, elegantly told.

What we see with this project is a visualization that answers "data questions". Data questions are the lines of interrogation and the dimensions of interpretation users will likely seek to pursue when reading a visualization design.

It's more than just framing a story, it is about the specific insights we are making accessible. It is the most defined and detailed level of editorial focus we should aim to achieve. We want our visualization to be able to respond to the most likely and relevant questions a user will raise about the data and the subject matter.

At this point, we are starting to consider the relationship between our editorial focus and the potential visualization design options.

As Amanda Cox describes earlier, the way that you choose to represent your data—the form you give it through your selection of chart type—should be influenced by the questions you are trying to answer.

For instance, if you are asking a chart to facilitate a comparison between the values of different categories, you might deploy a bar chart. You wouldn't use a line chart to achieve this, but you would if you wanted to show how a value or values change over time. The scatter plot we just saw was the perfect method of comparing two quantitative values for all those different countries. It was the right form to answer the specific data questions identified.

So we need to know what questions we're trying to answer.

Unless you've already had them specifically outlined to you, an effective approach to tackling this can be drawn from the practice of logical reasoning, specifically induction and deduction. These techniques are common to academic and scientific research.

Deductive reasoning involves confirming or finding evidence to support specific ideas. It is a targeted and quite narrow approach concerned with validating certain hypotheses. A deductive approach to defining your data questions will involve a certain predetermined sense of what stories might be interesting, relevant, and potentially available within your data. You are pursuing a curiosity by interrogating your dataset in order to substantiate your ideas of what may be the key story dimensions.

Inductive reasoning works the opposite way. It is much more open-ended and exploratory. We're not sure precisely what the interesting stories might be. We use analytical and visualization techniques to try and unearth potentially interesting discoveries, forming different and evolving combinations of data questions. We may end up with nothing, we may find plenty—the insights we observe may be serendipitous as we follow our nose for the scent of evidence. Fundamentally, this is about using visual analysis to find stories.

For most visualization projects, if we have the time, ideally we would seek to use both deduction and induction in conjunction in order to learn as much as possible about what stories the dataset can reveal about the given subject matter.

Using visual analysis to find stories

The following is a quote from Ben Schneiderman:

> *"Visualization gives you answers to questions you didn't know you had."*

In the *Chapter 2, Setting the Purpose and Identifying Key Factors*, we discussed the different intentions and motives you might have for developing a data visualization. In most cases we think of it as something we create and provide to others. What we sometimes neglect to consider is the potential of visualization for ourselves, when we are the intended users looking to discover insights about a subject.

This is where we consider the application of visual analysis. Visually analyzing a dataset, and employing both inductive and deductive reasoning, enables us—as the designer—to learn more about our subject by exploring a dataset from all directions.

As Ben Schneiderman articulates above, and as we saw through the demonstration of Francis Anscombe's experiment, rather than just looking at data, we are using visualization to actually see it, to find previously undiscoverable properties of our raw material, to learn about its shape, and the relationships that exists within.

This activity can also be described as data sketching or preproduction visualization. We are using visualization techniques to become more intimate with our raw material and to start to form an understanding of what we might portray to others and how we might accomplish that.

Visual analysis requires a high degree of graphical literacy, the ability to read and interpret data represented visually. This is something we might not really think about too often. In fact, if we're honest, many of us would probably have to admit that we can actually be quite passive in how we engage with a visualization or infographic.

This activity requires a much more committed level of attention to interpretation. As we explore the evolving visual analysis of our data, we need to be prepared to observe the following characteristics that will lead to the identification of our key stories:

Comparisons and proportions:

- **Range and distribution**: Discovering the range of values and the shape of their distribution within each variable and across combinations of variables
- **Ranking**: Learning about the order of data in terms of general magnitude, identifying the big, medium, and small values.

- **Measurements**: Looking beyond just the order of magnitude to learn about the significance of absolute values

- **Context**: Judging values against the context of averages, standard deviations, targets, and forecasts.

  Using methods like a bar chart will enable comparison across values and categories to pick out the type of physical qualities just listed, as shown here:

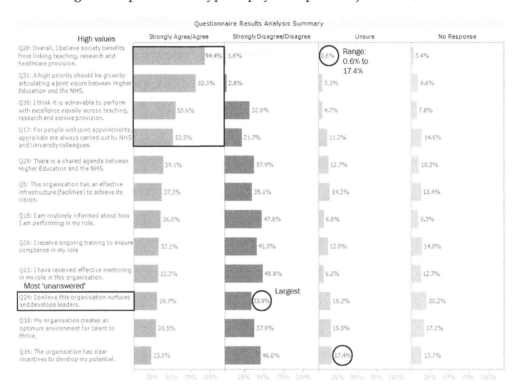

Trends and patterns:

- **Direction**: Are values changing in an upward, downward, or flat motion?

- **Rate of change**: How steep or flat do pattern changes occur? Do we see a consistent, linear pattern, or is it much more exponential in shape?

- **Fluctuation**: Do we see evidence of consistent patterns or is there significant fluctuation? Maybe there is a certain rhythm, such as seasonality, or perhaps patterns are more random

- **Significance**: Can we determine if the patterns we see are meaningful signals or simply represent the noise within the data?

- **Intersections**: Do we observe any important intersections or overlaps between variables, crossover points that indicate a significant change in relationship?

  Using a line chart is a perfectly suitable method to observe patterns and trends, as we see below:

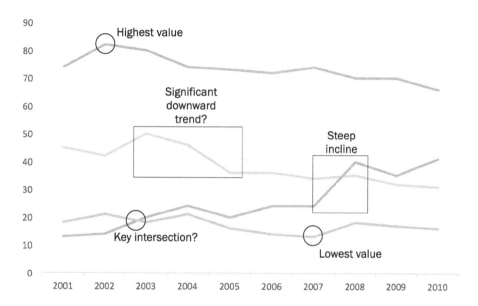

Relationships and connections:

- **Exceptions**: Can we identify any significant values that sit outside of the norm, such as outliers that change the dynamics of a given variable's range?

- **Correlations**: Is there evidence of strong or weak correlations between variable combinations?

- **Associations**: Can we identify any important connections between different combinations of variables or values?

- **Clusters and gaps**: Where is there evidence of data being "bunched"? Where are there gaps in values and data points?

- **Hierarchical relationships**: Determining the composition, distribution, and relevance of the data's categories and subcategories.

Using a scatter plot will enable visibility of these types of relationships, as shown below:

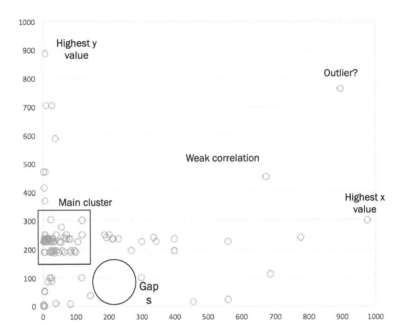

Through embarking on such in-depth visual analysis we should achieve the level of data intimacy required to refine our editorial focus. The visual interrogations we perform on the data will unearth evidence of the features listed over the previous couple of pages.. Where we find these, we will find the stories.

The process of visual analysis can potentially go on endlessly, with seemingly infinite combinations of variables to explore, especially with the rich opportunities bigger data sets give us. However, by deploying a disciplined and sensible balance between deductive and inductive enquiry you should be able to efficiently and effectively navigate towards the source of the most compelling stories.

The chart types that we have seen being used previously are illustrative of just a small section of the gallery of options we have to call upon. We will learn much more in *Chapter 5, Sketching, Planning, and Conceiving Your Visualization*, about the different chart types and their functions to understand which ones are best deployed for different enquiries of our data.

The product of our work here is a more sophisticated understanding of the stories existing in our datasets about the given subject matter. This will help us form the specific data questions that we'll be asking our visualization designs to answer. We've found our stories, now we need the appropriate methods to tell them and that's what *Chapter 4, Preparing and Familiarizing With Data*, will explore.

An example of finding and telling stories

Before we move on, to help embed the understanding of data familiarization, visual analysis and the difference between finding stories and telling stories, let's work through a basic example.

Take the following sample table of data. The subject matter is the Olympic games and specifically the total medals won by the top eight participating nations over five recent events. The selection of the top eight is based on them being the top ranked countries at the Beijing Olympics in 2008.

Suppose you were briefed to unearth some key stories around Olympics medal winning trends in recent years, how would you go about it?

Country	Total medals won in the Summer Olympics				
	2008	2004	2000	1996	1992
United States of America	110	103	92	101	108
People's Republic of China	100	63	59	50	54
Russian Federation	72	92	88	63	112*
Great Britain	47	30	28	15	20
Australia	46	49	58	41	27
Germany	41	49	56	65	82
France	40	33	38	37	29
Republic of Korea	31	30	28	27	29
ALL	951	929	925	842	815

* When part of former Soviet Union. Data from http://www.databaseolympics.com/index.htm

Let's start by just scanning the data with our eyes to find anything that stands out.

The main data issue appears to be that the Russian Federation medals total for 1992 was actually when it was known as the Soviet Union. It is noticeably higher than for all the other Olympic events, due to the contributions of additional member states that then made up the Soviet Union but who are now independent countries competing in their own right. As it will be hard to unpick this value to isolate just those athletes who would now be considered part of the Russian Federation, it will be sensible to just ignore this value from our analysis. Otherwise, it will skew our interpretations.

We can see that the event order goes from left to right in reverse chronological order and the vertical sorting is organized by the most successful nations as at 2008. In addition to the medal winning totals for the selected countries, we also have the aggregate of all medals across all countries.

We now continue our examination by noting some of the dataset's descriptive and statistical properties to develop an increased level of familiarity:

- Two variables: Country and event year
- Country is a categorical nominal variable with nine values (each country and the aggregate)
- Event year is a quantitative (interval-scale) variable with five values
- The maximum country medal count value is 110 medals, the minimum is 15
- The maximum aggregate value is **951** and the minimum is 815 (but that includes the Russian Federation contribution)
- Each event year is spaced 4 years apart
- The longest country name is People's Republic of China, the shortest is France

This gives us a sense of the physicality of the data and the potential influencing attributes that might shape our visualization architecture.

What other data preparation tasks might we undertake?

We have no real transformation activities to undertake in terms of addressing data quality aside from already deciding to ignore the Russian Federation total.

For transforming the data for its use in analysis we may decide to create some calculations to show the percentage of medals won out of each event total. You may also decide to abbreviate some of the county values to potentially help accommodate the space required for labeling.

We also need to consider data consolidation. For the purpose of this demonstration, we are going to stick to our original dataset on its own but there could be many different options to enhance and contextualize this subject matter, including the following:

- The details behind the medal totals of how many golds, silvers, and bronzes each country has won
- The full dataset of medal statistics for all the other countries who have competed, not just the recent top eight
- The full dataset of medal statistics for every Olympic games
- The number of competitors who were taking part in the games for each country, in order to understand the percentage of success of each team
- The split of performances between the different sporting events

- Population figures to contextualize the achievements, maybe even sporting participation figures if they were recorded
- Historical milestones of socio-political and geo-political issues to help us appreciate the status of the different countries at these key points in time
- You might look to bolster the ingredients of your visualization design resources with national flags' image files or URL links to national Olympic associations

Whether we could obtain these additional data items is another matter and they may not even help with our stories. But it is always good to let your imagination roam and explore ideas for content that could really enhance your work.

Our data is now in good shape. Next up, we look to develop our editorial focus, specifically considering the following:

- What initial sparks of curiosity crossed our minds when we were given the brief and initially saw the data?
- What dimensions of analysis do we think might be of interest or relevance about this subject matter?
- What data questions will we seek to answer in our visualization design?

To refine our focus we need to commence our visual analysis work to explore our dataset and see what comparisons, trends, patterns, and relationships we can identify. Out of this we will hope to unearth some interesting stories to tell.

Given we have a small dataset with only two variables we shouldn't need to embark on too much varied visual analysis.

The first graphic takes a look at the variation of medal winning across the years, showing the range of totals for each country using a floating bar chart:

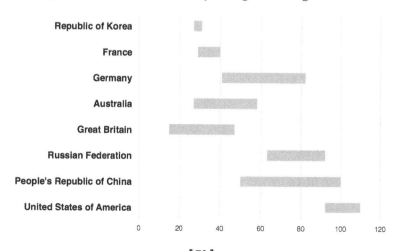

Through interpreting this chart in conjunction with the descriptive statistics we just collected, we are able to form some interesting data questions about the subject and start to get a feel about the main stories, such as:

Question	Answer
Which countries have experienced a significant change in their medal-winning performance levels?	We're looking for the widest bars to show the variability, this could be improvement, decline or inconsistency. We would identify the spread of Germany and China as being particularly interesting.
Which countries have maintained consistency in their performance levels?	Now we're looking for the narrowest bars, the tightest of value ranges. This leads to noticing the USA, France, and especially Republic of Korea.
What have been the most interesting country stories in terms of the transition of their performance and rankings?	Possibly too hard to see with this chart, but there is potentially something going on with the bars that intersect and exceed the lengths of others. At this stage, the story of China seems to stand out as being something to look out for.

Let's now repeat the same chart type but apply it to a transformed version of the data that has been standardized to show the medals won as a percentage of the overall total:

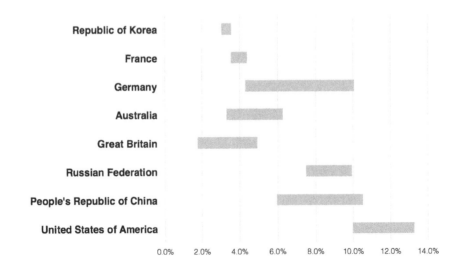

Does this alter the focus of our questioning or change our impressions of the main insights?

If anything it reinforces them, especially our interest in the varied performance levels for Germany and China. It also emphasizes the remarkable consistency of Republic of Korea and France.

At this point, we have definitely established a scent for the story. We have started to articulate the data questions that best interrogate this data and most likely reflect what the readers of a visualization about this subject will wish to learn.

We now need a different visual representation. Using the floating bar we have seen the categorical view of the countries and their performances. Now, we need to switch our perspective to the other main variable, that of event year, to pursue our curiosities about the transition of medal-winning performances and the transition in ranking of the individual countries across the five Olympic Games.

For this next visual sketch we turn to a line chart. On this single chart we plot the eight countries, differentiated by color, showing the absolute medal wins from left to right across the five Olympic events:

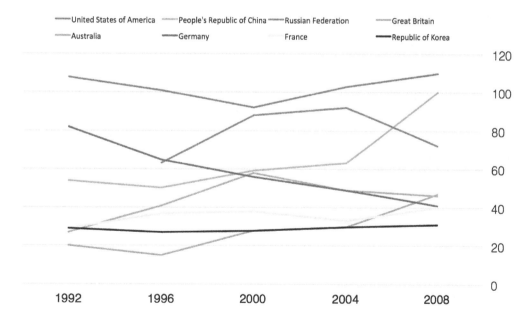

It looks a bit messy doesn't it? Don't worry. Remember, this is an exploratory visualization for ourselves. We are the audience and we just want to see if we can discover some interesting physical properties about the data in this display.

You wouldn't and shouldn't publish an isolated, cluttered, and poorly-annotated chart like this to convey a story to others, but when it is a visualization serving yourself, it is a different matter. You created it and you know what you're looking out for. Quick and dirty is absolutely fine.

The decision to place all countries onto one graphic is to enable visibility of the interesting transitions, the crossovers, the seemingly cluttered parts, and the empty parts. You could separate each country out into its own line chart and assess a matrix of eight small-multiples. However, this would only show you the individual country stories. Our keen interest here is in the relationship between the countries.

The chart shows how Germany's (blue) wide range of results, actually reflects their general decline in medal winning levels and, by extension, their relative rank.

By contrast, China's wide distribution shows a country on the rise over the past four games at least. The extended fascination of this trend would be whether they will catch up and possibly overtake the US once we have the results and data for the 2012 Games (not available at the time of writing this book!).

Elsewhere, Russia can be seen to have moved up and down over the years and has now been overtaken by China. There is an interesting chunk of white space for the 2008 results either side of the Russian value, leaving them quite comfortably in third position. Interestingly, the UK has seemingly demonstrated a very similar pattern of improvement relative to the Chinese over the past five events.

Sometimes no change is as interesting as some change and, in this respect, the consistency of Republic of Korea is quite stark given the different generation of competitors who will have contributed to those totals.

Otherwise there is nothing else really of significant interest. The charts have served their purpose in discovering and confirming some relevant and interesting stories concerning the contrasting experiences of China, Germany and, potentially, the Republic of Korea.

Of course, sometimes you simply may not find a story. There just might not be anything of substance to convey to others visually, in which case a table of data may prove to be the most appropriate solution.

However, we *have* found our stories, so how do we tell them? As a bridge to the next chapter, where we will be focusing on design matters around presenting our stories, let's attempt a quick solution.

Remember the quote we saw earlier from Amanda Cox: "different forms do better jobs and answering different questions"? Let's reduce the story to a simple contrast between China and Germany. Our main data question will be something like "how have the medal-winning performances of China and Germany compared over the past five events?"

The most suitable method for giving form to and answering this question will still be a line chart. Similar to the one we used for the visual analysis, we are trying to show the relationship between these two countries' respective performance over time.

However, the design execution will be different. This time we're conveying the story to others, so we need to refine the visuals in order to make it an explanatory piece:

- We need to elevate the important features of the main story and relegate any background context and secondary content.
- We need to ensure that there are annotations for labels, values, and captions so the reader is entirely clear about what is being communicated.

Here is a proposed solution for telling this story:

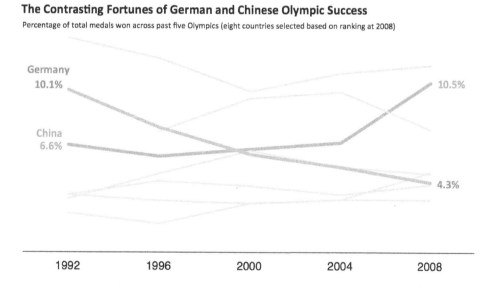

The Contrasting Fortunes of German and Chinese Olympic Success

Percentage of total medals won across past five Olympics (eight countries selected based on ranking at 2008)

The first thing to point out is that we have used the calculated data for medals won as a percentage of the total. This is more appropriate for this story as it helps standardize and contextualize the performance across all events in a more comparable way.

The aim here is to provide a clear visual hierarchy emphasizing the two main countries in our story and diminishing the contextualizing six nations into the background. We could have removed the other six countries but, through the use of a subtle shade of grey, we can still see them well enough to get a sense of the overall rankings. That is all we need from them—context.

The title neatly frames the story, the subheading describes the chart and the data, and the labels help the reader compare the two countries' relative trajectory.

The use of color attempts to help imply the positive improvement (orange = hot = good) of China and the negative decline (blue = cold = bad) of Germany. Only the bare minimum chart apparatus (the axis line) is included, once again, to allow the main story to come to the fore.

Contrast this design approach for telling a story (explanatory) with the design of the same chart method we used to find the stories (exploratory); here we provide nothing more and nothing less than the reader requires to easily interpret the story. This use of contrasting visual approaches for the same chart types but for different intentions is important to recognize in your design work.

Summary

In this chapter, we have learned about the importance of editorial focus and content reasoning—the ability to recognize the most important and relevant stories in your data and the discipline of taking responsibility to optimize the interpreting experience of your intended audience.

We worked through the mechanics of acquiring, preparing, and familiarizing with your dataset. In particular, we highlighted the importance of our own graphical literacy in the task of conducting visual analysis.

We identified numerous physical characteristics of our data that will help us to discover key stories and help inform the types of data questions we may seek to answer in our designs.

Finally, we worked through a demonstration of using visual analysis to make sense of your data, to find stories yourself and then tell those stories to others. We also saw an example of the contrasting visual design solutions used for exploratory and explanatory visualization.

Now that all our preparatory work has been covered, in the next chapter, we will move the methodology forward onto the design stage. Here we will learn about all the options we need to judge and the decisions we need to take across the five key layers that form the anatomy of any data visualization design.

4
Conceiving and Reasoning Visualization Design Options

So far the focus of our attention has been on the uncelebrated but vital preparatory and scoping stages of the data visualization methodology. We have established the purpose of our design and the key factors surrounding the project have been identified and weighed-up. We have also acquired and prepared our data and begun exploring it to identify the key data stories and analytical slices around which we may form our editorial focus.

These contextualizing activities are often neglected because they are understandably deemed not as fun as the design stage. Yet, they will save you time and pain, helping your work to proceed more efficiently by avoiding blind alleys and creative misjudgments.

In this chapter, we will be taking a forensic look at the many design choices involved in the process of establishing an effective visualization solution. We will tackle these choices by working through the anatomy of a visualization design, separating our thoughts into the complementary dimensions of the representation and presentation of data.

For rookie and experienced designers alike, the framework of design considerations outlined here should help you successfully navigate through the creative challenges and opportunities you are facing.

Data visualization design is all about choices

Here is a quote from Jer Thorp (`http://blog.blprnt.com/blog/blprnt/138-years-of-popular-science`):

> "*My working process is riddled with dead-ends, messy errors, and bad decisions — the final product usually sits on top of a mountain of iterations that rarely see the light of day.*"

Over the past two chapters, we have explored some of the key preparatory activities of the visualization design process. In doing so we have built a detailed level of clarity about what we want to achieve in our visual communication and why.

The heading for this section may seem obvious but it needs stating. As we'll see in this chapter, the scope for creativity can be quite overwhelming. How well you rationalize the many decisions you face throughout the process will strongly determine whether you achieve an effective visualization design.

To frame this discussion, do a quick image search in Google for the term "data visualization" and scroll through the first few screens. You will see just a snapshot of evidence of the innumerable variety of ways in which you can represent data. Some are good, some are bad. Some are really bad. Some shouldn't even be connected with the term data visualization.

Additionally, have a glance through the collection of submissions uploaded on to sites that run some of the main data visualization design contests (for example, `www.visualizing.org`, `www.infobeautyawards.com`). Choose a particular contest and explore the spectrum of proposed solutions, all typically emerging from the same dataset and responding to the same brief.

What can we learn from this? What does this evidence of the variety of ways in which people interpret visualization design challenges reveal to us?

The first thing to say is that there is never a single path towards a "best" solution. The inherent creativity and individualism of design work ensures that. An idealistic desire for a single and simple set of rules to achieve a guaranteed effective solution is simply unreasonable due to the many different factors that will shape the scope and intention of any given project.

There is, however, an established body of theoretical and practical evidence that guides us to understand which techniques work better for certain situations and less well for others. Importantly, these guides transcend instinct or personal taste and help us frame many of our design options, influencing the choices we make.

Beyond that it is more about managing trade-offs, about trusting your judgment to make sense of the problem context in which you are working, the requirements you are responding to, and keeping in mind the overall objectives of visualization design, as outlined in *Chapter 1, Context of Data Visualization.*

The second key observation is to remark that the very moment we take on a visualization challenge, and start our journey towards a design solution, we are commencing a unique creative route formed by numerous permutations of choices. Nobody else will go through the same experience nor arrive at exactly the same solution.

You won't always get there easily. That's important to recognize too. As Jer Thorp expressed in his quote, even the best make mistakes and end up wasting time following ideas that lead nowhere and having to change course halfway through. However, by following the approach we have outlined in this book, and specifically the framework of considerations for this chapter, we hope to reduce the waste and eliminate inefficiency. This allows us to fail faster and recover more quickly.

A useful way to look at a data visualization challenge is to recognize that we are actually seeking to reduce choices. This is achieved through recognizing influential factors, by considering the desired function and tone of our work, familiarizing with our data and identifying stories. We are building clarity through selection and rejection. We are reducing the problem by enhancing our clarity.

The reasoning involved in eliminating options is just as important a skill as determining those we shall pursue. This lets us control our work, it helps us plan better, and prepare for the creative avenues down which we may proceed.

In many ways you could equate this design process with the responsibilities of being a film director, managing the dramatic, artistic, and technical aspects of a film. A director has to create the film's vision, direct the cast, manage the crew, oversee the script, coordinate the choice of locations, the music, and the post-production effects. All these different perspectives require separate attention and unique treatments until they are brought together into a cohesive single product: a movie. We're trying a similar approach with our visualization design.

As we now move into the creative stage it is helpful to follow a framework that will help you to understand the many different design options about which you will have to make a decision.

An effective way to think about this is to consider the visualization "anatomy". By definition, anatomy refers to "the bodily structure of an organism", so we are appropriating the term to apply it to the structural layers of a data visualization design.

In the first chapter, you will recall the proposed definition of data visualization and how it separated the idea of representation and presentation of data. We see these as separate dimensions of our design task:

- **Data representation**: This is the foremost layer, how we give form to our data through the use of "visual variables" to construct chart or graph types.

- **Data presentation**: This is the delivery format, appearance, and synthesis of the entire design. It concerns the layers of color use, interactivity, annotation, and the arrangement of all elements.

Some helpful tips

Before we commence our design thinking, here are a few useful tactics to help you move smoothly through this process and achieve the best outcome:

- **Sketching**: Drawing out rough ideas on paper is a really good discipline to get into before you go anywhere near a computer. It doesn't matter if it is on the back of a beer mat, on a white board, or in a beautiful portfolio pad, try to sketch out your thoughts and concepts in order to download those ideas floating around in your mind. This is shown in the following screenshot:

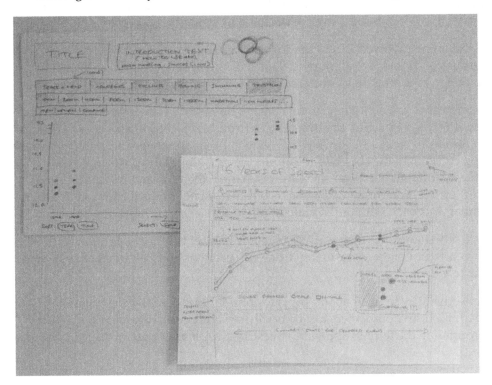

This is especially invaluable if you are collaborating with others and the creative process is a shared experience. Doing this could save you a lot of time clarifying and formalizing your collective thoughts. It will also be a safeguard against the risk of forgetting those great initial sparks of creativity that came to mind when the project was first triggered. It will also enable you to quickly refine and change direction, if necessary, without having committed time to any technical development.

- **Note taking**: As you develop your expertise in data visualization, one of the best methods to improve your creative judgment is to document your decision-making process. Keep a log of all the choices you've made, the reasons why you've decided to do one thing or reject something else, the ideas you've had along the way. You should also record details of any procedures you have established for tasks such as cleaning or transforming data or the cropping treatment applied to a set of images—you may need to repeat these and follow the same exact stages later. Finally, it is vital to make a note of all your data sources and references (including things that have influenced/inspired you). All this material will be particularly beneficial if you have the opportunity to publish a narrative of your design process, which can be extremely educational as much to yourself as it is to others.

- **Time management**: As with any design project, the planning and allocation of your time and available resource is a vital discipline. Utilize time wisely otherwise you can quite easily get sucked into spending too long on the minutiae of one stage (especially on the more attractive creative tasks). Don't be surprised to find that a great deal of your time is spent doing relatively mundane tasks such as data cleaning or preparation. That is to be expected. Also, developing an awareness of your typical design activity duration is very important, especially if you are a freelance designer looking to develop a more sophisticated approach to estimating and pricing for clients.

- **Reinventing the wheel**: As the field continues to mature and innovative new techniques are developed, there is a constantly growing reference library of potential solutions. As you will see soon, we shouldn't always need to feel like we have to constantly invent something new. There are plenty of creative options already available to be influenced or inspired by, providing us with idea templates to develop and tweak for our needs. Of course, you should always attribute influence and reference designers' work when it has been utilized as a significant and apparent source of help.

The visualization anatomy – data representation

The process of identifying the most effective and appropriate solution for representing our data is unquestionably the most important feature of our visualization design. Working on this layer involves making decisions that cut across the artistic and scientific foundations of the field.

Here we find ourselves face-to-face with the demands of achieving that ideal harmony of form and function that was outlined in the objectives section of *Chapter 1, Context of Data Visualization*. We need to achieve the elegance of a design that aesthetically suits our intent and the functional behavior required to fulfill the effective imparting of information.

What we're doing here is determining how we are going to show what it is we want to say. It is a difficult skill to master—something of a dark art—particularly given the set of factors we need to consider and the trade-offs we might need to make. Our task involves considering the following:

- Choosing the correct visualization "method" for the stories we're telling
- Accommodating the physical properties of your data
- Facilitating the desired degree of precision
- Creating an appropriate metaphor to depict our subject stylistically

Choosing the correct visualization method

The first matter is to determine the choice of visualization method. We aren't necessarily committing just yet to a specific chart or graph type, though we might have some in mind. Rather, this is about the general family or collection of chart types as defined by their primary storytelling method.

For example, a bar chart serves the function of comparing categories of values; a line chart, by contrast, enables us to show changes of values over time, geo-spatial data can often be best displayed over a map.

Your choice of visualization method will be mostly driven by the outcome of your work in *Chapter 3, Demonstrating Editorial Focus and Learning About Your Data*. You've developed your editorial thinking about the key data stories, analytical dimensions, and the questions you're trying to answer in your visualization.

Of course, it is often likely that you have determined several different analytical slices and you will probably need to consider different methods to appropriately convey the stories for each one.

There are a number of ways of classifying the variety of methods for visualizing data but here is a suggested taxonomy:

- Comparing categorical values
- Assessing hierarchies and part-of-a-whole relationships
- Showing changes over time
- Mapping geo-spatial data
- Charting and graphing relationships

Of course, there are often overlapping functional or storytelling features inherent to the chart types that sit under these method headings. For instance, a stacked area chart shows changes over time but also facilitates the categorical comparison of its different layers. That would be an example of a chart type that spans across two method classifications. However, the principle focus of this chart type is telling a story over time and so we would consider it belonging to the "showing changes over time" method. The comparisons it enables represent an additional but secondary focus.

As we saw in the previous chapter, the forming of data questions really helps you articulate the range of analytical stories you are wishing to portray. In our demonstration exercise, when we were looking to show the results of our analysis, we were essentially responding to the question "how have the medal-winning performances of China and Germany compared over the past five events?"

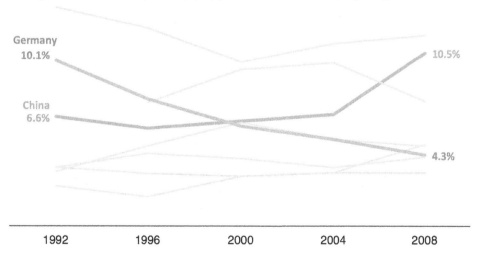

The Contrasting Fortunes of German and Chinese Olympic Success

Percentage of total medals won across past five Olympics (eight countries selected based on ranking at 2008)

The story being expressed was about showing changes over time: that defines our method. The selection, ultimately, of a line chart—a type belonging to this category of visualization methods—was evidently a suitable match as we specifically wanted to show the detail of the continuous transition across all five event years.

Had the focus been more about a comparison of all countries and the combined, aggregate picture of results over time we might have chosen a stream graph or an area chart. Both of these are chart types that would typically fall within the method of "showing changes over time".

Alternatively, had we sought to demonstrate the stark comparison of the medals won at the earliest and most recent events in our dataset, we might have chosen a method for comparing categorical values. In this case, the use of a slopegraph or a bar chart would have been more suitable, as shown here:

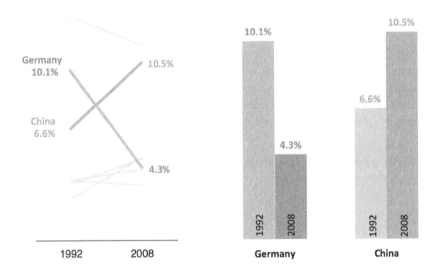

We will cover this issue in more depth in *Chapter 5, Taxonomy of Data Visualization Methods*.

Considering the physical properties of our data

Now, we're looking to narrow down our search further by thinking about which types of charts will most effectively accommodate the variables of data we're looking to portray.

As discussed in the previous chapter, learning about the physical properties of your data gives you an important sense of the shape and size of your data. As you refine your editorial focus you will develop an understanding of the data variables you may seek to display graphically.

The quantity and nature of the variables you are using will have a significant influence on reducing the range of suitable chart types you might be able to use within the method family you have chosen. As discussed earlier, this process of eliminating choices can only be of help to us as we move forward.

Referring back, once again, to our demonstration for the Olympics project, the data we were looking to use for our final story was event year (quantitative interval-scale), medal totals (quantitative ratio-scale), and country (categorical nominal). We had a good sense of the range and distribution of values held against each variable, we were just highlighting two countries and we wanted to show the full five-event transition. The best solution, therefore, was to use the line chart as we have just seen.

In *Chapter 5, Taxonomy of Data Visualization Methods*, we go into much more detail about this taxonomy and the range of chart types that sit underneath each of these five headings. You will see a gallery of some of the most common, contemporary methods. Each example is accompanied by a description of the chart and an outline of the data variables that each one can realistically accommodate. This will give you a good sense of some of the common data representation techniques. It could act almost like a creative menu for you to refer back to when seeking ideas and potential solutions.

Determining the degree of accuracy in interpretation

Now we start to step into the minefield. You can be assured that this section will have been the most revised and rewritten across the entire book.

Having identified the general visualization method and started to filter down further to identify the most suitable chart types, we now have to consider another key issue.

This judgment gets to the very heart of the form/function or art/science fault lines that exist in this field — to what degree of accuracy do you wish readers to be able interpret values from your visualization?

You might ask in response, why would you *ever* not wish to maximize the precision of interpretation? Surely, the mission is to deliver as much accuracy through our representation as possible?

Well yes, of course we do, but the inclusion of terms such as "maximize" and "as possible" allude to the specter of alternative influences. You see, for certain contexts, as we outlined in the early chapters, you might be seeking to explore different aesthetic forms of representation. And yes, sometimes this might involve certain sacrifices in terms of the precision of interpretation.

To frame this section, we first need to learn about visual variables. A visual variable is the specific form we assign to data in order to represent it visually. It could be the length or height of a bar, the position of a point on an axis, the color of a county on a map, or the connection between two nodes in a network.

Each of the chart types that we come to take as common representation methods are based on the deployment of a single or, more commonly, combination of several visual variables at once. Using multiple variables, in particular, enables a designer to efficiently express extra layers of meaning behind the properties of a single mark, as the next example demonstrates.

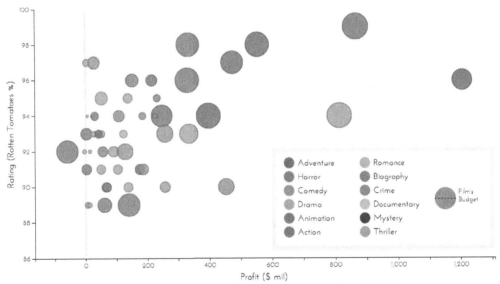

Image from "How Much Money do the Movies we Love Make?"
(http://vallandingham.me/vis/movie/), created by Jim Vallandingham

In this bubble chart visualization, each mark involves a combination of several visual variables representing a range of different data variables:

- The position on the x axis represents a film's profit.

- The position on the y axis represents the average review percentage rating of each film.

- The circle area represents the film's budget.

- The circle's color (hue) represents movie genre.

- Users interacting with this web-based design will also discover a text label displaying the raw values by hovering over one of the bubbles. Text is not universally treated as a visual variable but it is still worth acknowledging.

If you let your imagination run free and try to conceive as many visible properties as possible that might be capable of representing series of categorical, ordinal, or quantitative data, you will realize that there are many potential approaches. In fact, in a fun recent experiment (`http://blog.visual.ly/45-ways-to-communicate-two-quantities/`), visualization designer Santiago Ortiz proposed over 40 different ways to potentially represent just two simple numbers.

Indeed, if you fully release your creative inhibitions, you can take things beyond the visible or physical, as some of Santiago's suggestions did, and consider how our other senses might be exploited in order to represent/interpret data through channels such as sound, smell, touch, and taste. Just imagine how some of the inherent variable qualities of our other sensory mechanisms could be capable of distinguishing categories and values of data.

Anyway, back to more practical thoughts, for now, and a brief history lesson.

Once upon a time there was a man called Jacques Bertin. Bertin was a clever chap. In fact he was a pioneering thought-leader within data visualization. Building on the earlier studies from the Gestalt School of Psychology — mentioned in *Chapter 1, The Context of Data Visualization* — his book *Semiologie Graphique* (1967) is one of the subject's founding texts and represents one of the earliest and most comprehensive attempts to theorize how we perceive and interpret different representations of data through shape, pattern, and color.

Bertin determined that there were three main aims behind your choice of data representation, moving from high-to low-level acts of graphical interpretation. This is still an extremely potent way of organizing our thoughts and reasoning our selection of the most effective visual variables. These aims are as follows:

- The highest level of Bertin's interpretive acts concerned whether we are able to visually discriminate between different data marks or data series: can we actually see and read the data being presented. We must make sure that the way we visually distinguish different categorical and quantitative values is legible and is in no way hidden by way of unnecessary clutter, noise, or distraction.

- The second act refers to being able to satisfactorily judge the relative order or ranking of values in terms of their magnitude. This is basic pattern matching where we seek to determine the general hierarchy of the values being displayed: where is the most and where is the least, which is the biggest and which is the smallest.

- The lowest-level act relates to judging values. Studies have shown how the effectiveness of different visual variables can be ranked based on which most accurately support comparison and pattern perception. Bertin was the first to propose such a hierarchy and his work has been tested, developed, and refined by Cleveland and McGill (*Journal of the American Statistical Association, Vol. 79, No. 387. September, 1984, pp. 531-554*) and then by Jock MacKinlay.

Here, in the following presentation, we see the most recent version created by MacKinlay. Each column represents the three main data types (note that there is no distinction between ratio and interval-scale types of quantitative variables). Within each column you have an ordering of the most accurate and least accurate visual variables according to their interpretive precision:

	Categorical (Nominal)	Categorical (Ordinal)	Quantitative (All)
Most Accurate	Position	Position	Position
↑	Color Hue	Density	Length
	Texture	Color Saturation	Angle
	Connection	Color Hue	Slope
	Containment	Texture	Area
	Density	Connection	Volume
	Color Saturation	Containment	Density
	Shape	Length	Color Saturation
	Length	Angle	Color Hue
	Angle	Slope	Texture
↓	Slope	Area	Connection
	Area	Volume	Containment
Least Accurate	Volume	Shape	Shape

Image recreated from "Ranking of Perceptual Tasks" (*Automating the Design of Graphical Presentations of Relational Information, ACM Transactions on Graphics, Vol.5, No.2, April 1986*) by Jock MacKinlay.

The studies by Bertin, Cleveland and McGill, and then MacKinlay focus on the fact that our visual system isn't capable of absolute measurements. Therefore, frameworks like this simply propose a guide to understand which variables will be better at delivering relative measurements but with highest accuracy. In other words, the higher up the column the easier it will be for your reader to accurately interpret values represented by those variables.

So, looking at that table, you might ask why you would ever *not* use position as the visual variable for your data: That will surely maximize the efficiency and accuracy of your data communication for all data types?

It is unfortunately not as simple as that. If it was, I don't think I would need to write this book. As we've seen earlier, we rarely only have just one data variable to communicate. You will therefore often need multiple visual variables to communicate multiple data variables.

However, above that, and returning to our section introduction, how accurate do we really need the interpretations to be? Do we actually need to facilitate the reading of exact values from every visualization we create?

Alternatively, can we allow ourselves more creative freedom by recognizing that in some cases just being able to facilitate the relative order of values may be sufficient for the context and requirements of the design?

If you recall in the first stage of our methodology we discussed the importance of trying to define, as early as possible, the functional and tonal quality of your intended design. The tonal judgment, in particular, is the important matter right now for this is what separated those pieces that matched an analytical and pragmatic style from those that were more abstract or emotive.

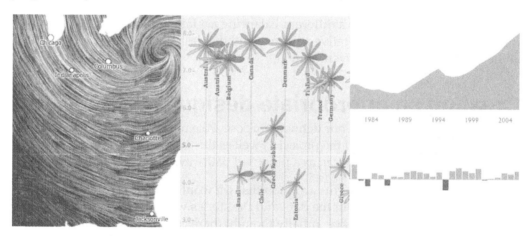

In this image we see a selection of visualization styles and demonstration of the fine balance being judged between design creativity and interpretive accuracy based on the contextual requirements. Let's take a closer look at each one at a time:

- The "wind map" on the left-hand side of the image doesn't aim to facilitate the reading of exact values. The use of pattern density to indicate the strength of the wind, as shown in the hierarchy table, focuses on delivering a sense of those areas with strong wind (as well as its direction)and the areas where there is little wind. The elegance of the resulting design makes for a compelling visual that draws users to interact and learn about the patterns.

- The "OECD Better Life Index" display in the middle shows a glyph chart based around floral shapes with the petals sized according to different quality of life-indicator values. We don't find it that easy to determine precise values from these shapes but we do get a sense of the big values, the medium values, and the small values. This is an attractive alternative to a very pragmatic and analytical display such as a bar chart, or even just a table of numbers. At this primary layer of the tool's interface, the balance achieved between design creativity and interpretive accuracy was judged to be ideal, with the added feature of interactivity to enable more detailed annotation and accurate value-reading.

- The example of an area chart and bar chart, on the right-hand side of the image, provides a contrasting context. Here we might be talking about an analytical experience where the accuracy and efficiency of exchange is paramount. The idea of design innovation or novel creativity is not important. In these cases, you will be looking to prioritize the deployment of the higher-ranking visual variables enabling a reasonably easier experience in reading the values.

In each of these cases, we see a different balancing act taking place, a series of trade-offs between the interpretive accuracy and the design aesthetic to arrive at the right solution for the given context.

Creating an appropriate design metaphor

Maintaining consistency with our defined purpose—the requirements that come from understanding what triggered the project as well as the tonal and functional choices behind our intent—should be seen as a proposed pathway, not a final destination.

It may be that our initial thoughts around what we would hope to achieve are revised as we learn more about the data and the stories we can and may wish to tell. For instance, we might have initially thought there could be a rich narrative emerging that could have been portrayed quite powerfully and emotionally. As you learn more about your data and its potential deployment representation-wise, it may be that you realize a more analytical approach is more suitable.

Likewise, when we discover more about the extent of variety and potential in a dataset, a topic we thought would lead to an explanatory piece may actually evolve to be a more exploratory piece.

We are never fixed to our choices, but the quicker they are defined and the clearer we are in our conviction the better the design will be served.

This is important to recognize because it may be that when you start compiling your potential representation solution, you start to see things in a different light. This is particularly possible when the designer starts to consider integrating an extra layer of meaning through visual metaphor.

Visual metaphors are about integrating a certain visual quality in your work that somehow conveys that extra bit of connection between the data, the design, and the topic. It goes beyond just the choice of visual variable, though this will have a strong influence.

Deploying the best visual metaphor is something that really requires a strong design instinct and a certain amount of experience.

In this next example, designer Moritz Stefaner was commissioned to analyze and visualize how the clients and customers of a German start-up muesli company combined the ingredients they offer. The end result was a static visualization based on a radial network or chord diagram (on the right), showing ingredients grouped by category (base mueslis, fruit, nuts, sweets, and so on):

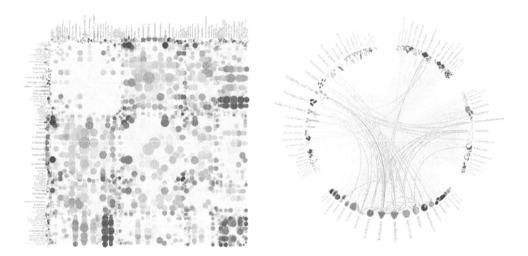

Images from "Müsli Ingredient Network" (http://moritz.stefaner.eu/projects/musli-ingredient-network/), created by Moritz Stefaner for mymuesli (http://mymuesli.com)

When describing his design process, Moritz noted how, when he worked on different sketches of the data, the matrix chart (example on the left-hand side) revealed some particularly strong stories that were otherwise missed in the radial diagram (on the right-hand side).

However, the radial diagram was the solution picked for the final piece. So why was this? Aside from some printing-readability issues that undermined the matrix chart, he commented: "from a visual point of view, it does not look very tasty." The radial diagram looked more appetizing—more edible—and fitted more strongly with the metaphor of a fruit bowl. The matrix chart solution looked more like fungi!

The decision to sacrifice certain qualities of interpretive enlightenment—offered by the matrix chart—was justified by the designer's instinct to enhance the visual metaphor emerging from the design of the radial diagram.

Referring back to the "wind map" again, here we saw a visual design that seemed consistent with the subject matter: it looks and feels like wind would look like in our imagination. In the piece that we saw in *Chapter 2, Setting the Purpose and Identifying Key Factors*, titled *Iraq's Bloody Toll*, this was a powerful story made even more impactive through the subtle but emphatic visual metaphor created by the arrangement of the chart and the color scheme.

These examples reinforce the value of Edward Tufte's message, from earlier, about the importance of making judgments through our own lens and based on our own word of authority.

Choosing the final solution

From the options and influences that we have just processed, we should be able to narrow down and identify the right data representation specification for our visualization.

It may be that this specification matches neatly with an established chart type and we can essentially "pick it off the shelf" and deploy it. This could be considered our top-down approach. In *Chapter 5, Taxonomy of Data Visualization Methods*, we'll see a gallery of contemporary solutions. From this selection you may identify a particular solution that fits with your context both structurally and metaphorically.

Alternatively, we may wish to custom-build a solution from the bottom up, carefully constructing a design one visual variable at a time, accommodating the range of data variables we want to show, and the style with which we want to show it. Of course, every chart type had to originate from somewhere and be invented at some point. It may be that as we construct our unique solution, we end up slipping back towards realizing that a tried-and-trusted option remains the best choice.

As I introduced at the start of this section, getting the right data representation solution can be hard, particularly in light of all these competing influencing factors. The best way to enhance your skill on this front is through practice, developing your experience, and learning from others. Get into a discipline of curating great examples that you come across in the field and try to discover how others have tackled similar subjects or maybe even similar dataset "shapes".

The fundamental challenge is being able to handle the temptation (or pressures, depending on your viewpoint) to focus on achieving aesthetic innovation or novelty. This is understandable. Many times I hear clients and training delegates expressing a desire to simply move beyond the bar chart, line chart, and pie chart in order to create something different.

We are such taste-driven beings; it is simply human nature but sometimes this is seeking innovation for the sake of it. There is no point pursuing something different on the flawed basis of just trying to stand out from the crowd or to put more "bums on seats" if the resulting experience for the reader is one of ultimate frustration. The disappointment caused by aesthetics that obstruct and obscure the discovery of insights about the subject matter will outweigh any good will created by an initially positive impression.

The key is not to set out to achieve an attractive and attention-grabbing work—let those qualities emerge as a by-product of good design. Focus instead on delivering the appropriate functional elements by employing the most suitable data representation.

Over time, the more experience you gather as a designer, the more natural these judgments will become.

The visualization anatomy – data presentation

The presentation of data involves thinking about pretty much every other design feature that might be included in our visualization. Here, we are determining the following:

- The use of color
- The potential of interactive features
- The explanatory annotation
- The architecture and arrangement

The decisions we make about these layers should be focused on delivering extra meaning, intuitiveness, and depth of insight to our readers or users.

One of the key concepts throughout our judgment of presentation-related design options is to seek to make the visible, invisible. In contrast to data representation, where our objective is to make the invisible stories and insights, visible, data presentation features should almost feel invisible so that the portrayal of the data maintains visual dominance. Therefore, try to bear the following two things in mind:

- **Visual inference means data inference**: If it looks like data, it should be data. If it isn't data then you've incorrectly conveyed a sense of representation where there isn't any and design refinement is required. An example might be the use of a color to represent a certain sentiment. If that color is used on a bar chart or is picked for the background of a label or call-out, but it is no longer connected to the representation of any sentiment meaning, this may trick the reader who has programmed their visual sense to spot this inference.

- **Facilitating the resemblance of data**: Let the data breathe. We talked about this in the discussion about Jacques Bertin's interpretative acts, but the presentation layers of your visualization will have a great impact on this. Ensuring a reader can discriminate between data categories and values is usually influenced by the background artifacts and surrounding apparatus. Throughout your design, make sure your data stands out clearly as the principle visual component.

The use of color

Here is a quote from Maureen Stone (http://www.perceptualedge.com/articles/b-eye/choosing_colors.pdf):

> *"Color used well can enhance and clarify a presentation. Color used poorly will obscure, muddle, and confuse."*

We've already touched on various aspects of using color as a potentially important visual variable for the representation of data, but the deployment of color for a visualization project naturally extends further. Given the depth and breadth of the field of color theory, it is important to consider it separately from our other design choices. The preceding quote emphasizes the value of doing this.

When deployed poorly, the use of color can create unnecessary decoration that can distract and compete undeservedly for attention in ways that will undermine the clarity and accessibility of the information exchange.

Conversely, with effective use of color we can deliver an attractive, synthesized design that most efficiently taps into the preattentive nature of the eye and the brain.

We are seeking to create layers of visual prominence that help us instantly achieve a sense of the important messages and features. Take a look at a landscape painting and witness the depth that is created through color, the separation between foreground and background that helps elevate prominent features and relegate contextual properties.

The best advice for guiding your decisions about using color is to refer to the two key rules shown at the start of this section—make sure it is used unobtrusively and it does not mislead by implying representation when it shouldn't be.

As with all design layers, the sensible objective here should be to strive for elegance rather than novelty, eye-candy, or attractiveness. To achieve this, it is important to be aware of the different functions, choices, and potential issues surrounding color deployment.

To represent data

One of most common mistakes used in relation to color is seen when it is being deployed to represent quantitative data. Specifically, when the "hue" property of color is used.

Take a look at this spectrum of colors: if these squares were representing quantitative data, which would be the biggest? How about the smallest? Which is bigger, red or blue?

As you will realize, there is no convention or association that determines a relationship between color (hue) and any sense of hierarchy or order of magnitude. We don't see one color as being inherently bigger or smaller than the other, and so to use this to represent quantitative data is a mistake.

In the following pair of images, on the left-hand side, even with a color legend explaining the value bands being depicted by the different colors, there is no preattentive association that allows us to efficiently determine the values being represented on the map. Referring back to Bertin's interpretive acts, we can't even easily establish a general sense of big, medium, and small values without having to constantly move to-and-fro the map and the legend. By contrast, the map on the right-hand side uses a single hue and uses a sequential color scheme that represents the highest values (dark) to lowest values (light) in a logical and immediately understandable way:

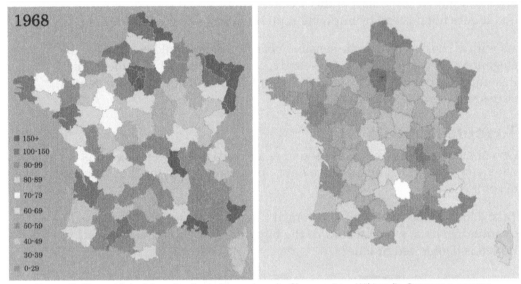

Image (left) republished from the freely licensed media file repository Wikimedia Commons, source: http://en.wikipedia.org/wiki/File:FrancePopulationDensity1968.png
Image (right) from "The Good and The Bad [2012]" (http://www.theusrus.de/blog/the-good-the-bad-22012/) by Martin Theus

What we can see demonstrated in this example is that, for quantitative data, one of the best ways to effectively depict a range of quantitative values is through the lightness property of color: that is, a scheme which goes from the most intense color through to increasing amounts of white. This is sometimes called a sequential color scheme:

As we can see clearly in this next display, we inherently and automatically attach a sense of order to such sequential scales. Of course, without a key it might be difficult for us to precisely pick out the absolute values that each color band represents, but we can certainly determine major patterns that lead to judgments of data order within and across both sample maps:

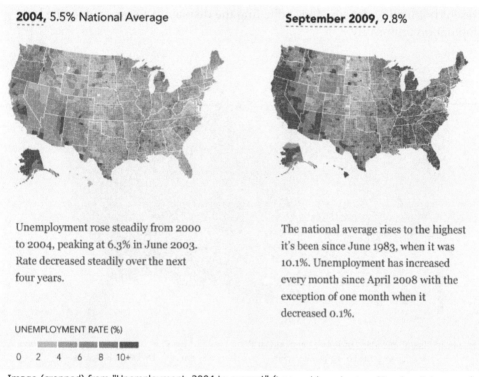

2004, 5.5% National Average

September 2009, 9.8%

Unemployment rose steadily from 2000 to 2004, peaking at 6.3% in June 2003. Rate decreased steadily over the next four years.

The national average rises to the highest it's been since June 1983, when it was 10.1%. Unemployment has increased every month since April 2008 with the exception of one month when it decreased 0.1%.

UNEMPLOYMENT RATE (%)

0   2   4   6   8   10+

Image (cropped) from "Unemployment, 2004 to present" (http://projects.flowingdata.com/ america/unemployment/raw.html), by Nathan Yau.

That idea, of surfacing the general patterns of the highest and lowest values, is really what the main purpose of color is when used to represent quantitative variables.

There are other types of color scheme used for situations that require us to represent two quantitative variables or to highlight two extremes of a single variable. These are known as diverging schemes.

While there is a variety of different ways to construct diverging color schemes, typically, the extreme ends of the spectrum are presented as darker and distinguished by strongly contrasting color hues. Alternative approaches might involve exploiting established color metaphors or might already be intuitively understood or easily learned.

The next image is an example of where preprogrammed understanding of color representation can be utilized. In this case, we see the respective strength of party political support across the U.S., with the Republicans represented by their established red and Democrats in blue. This is a topological map that displays calculated contours to show the general spread of support for each party. An added dimension to this particular piece is the use of an extra representation — color transparency — to represent population density, thus adjusting the display to accommodate the lack of population uniformity.

Image from "Isarithmic Maps of Public Opinion Data" (http://dsparks.wordpress.com/2011/10/24/isarithmic-maps-of-public-opinion-data/), by David B Sparks

It isn't just on maps, of course, where properties of color can be important to distinguish quantitative values. One of the most popular methods for coloring involves the traffic light metaphor of red, amber, and green. This is commonly used in corporate settings to indicate good, average, or bad performance thresholds.

However, it is important to know that around 10 percent of the population (particularly males) has a red-green color deficiency. The use of an approach such as the traffic light colors will therefore potentially alienate a significant proportion of your intended audience. An effective alternative is to switch green for blue, so positive values are now shown as blue and negative are still in red, as we see in the following horizon chart:

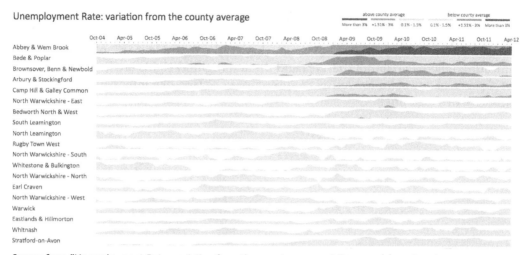

Image from "Unemployment Rate: variation from the county average" (`http://warksobservatory.files.wordpress.com/2012/07/unemployment-horizon-chart.pdf`), by Spencer Payne/Warwickshire Observatory

To check your chosen color schemes against the potential impact of different color deficiencies, use an application such as Vis Check (`http://vischeck.com`), which is a free online tool to simulate what a color-blind person would see when looking at your images.

As we've already explained in the data representation section, one of the key functions of a visual variable is to facilitate resemblance — the discrimination of data — and the use of color (hue) to distinguish between categorical variables is a particularly strong aid.

In the next example, we see a project created to display the status of various indicators surrounding how different states around the U.S. handle gay rights issues.

There are seven distinct categories of data distinguished by a unique color. The color itself has no meaning; it is purely a means of helping to separate out the various tracks of issues. The lightness of the color does add an extra layer of information, indicating where maximum (darker) and limited (lighter) rights are in place, and the absence of any color as well as the presence of a cross-hatching pattern further encodes extra meaning:

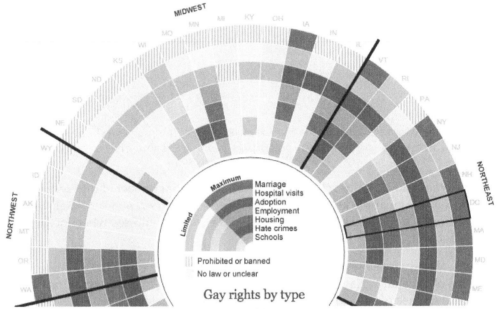

Image from "Gay rights in the US, state by state" (http://www.guardian.co.uk/world/interactive/2012/may/08/gay-rights-united-states), by Guardian in America Interactive

As we saw in the earlier image showing the political persuasion map of the U.S., the use of color for categorical data also allows us to maximize the implication of metaphorical or representative association.

However, regardless of whether the color depiction of categories is arbitrary or embodies more meaningful association, one of the key rules we need to obey is that the eye is only really capable of distinguishing up to a maximum of twelve different color classifications. This is just one of the many fascinating aspects of color that can be discovered from great books such as *Visual Thinking by Design*, by Colin Ware.

If you have more than twelve categories you may need to find ways of combining classifications to avoid this issue. You'll see this in effect on images such as the many subway maps around the cities of the world. As more extended lines and routes emerge, there are fewer remaining distinct color options that will help to emphasize, indicate, and separate these new markings.

There are also many definitions about the emotional or cultural significances behind color representation. It is naturally advantageous to exploit universal visual languages, but only when they are definitely universal! You need to be sensitive to the potential differing perceptions of color meaning across the regions of world. For most colors there is contradictory association and so referring to a resource such as *Color Meanings by Culture* (`http://www.globalization-group.com/edge/resources/color-meanings-by-culture`) will represent time well spent.

To bring the data layer to the fore

In addition to the representation of data, we also look to employ color to help create visual depth and a sense of hierarchy in our designs. In the first chapter, we saw the demonstration of color and imagery being at fault for the lack of clarity in the diagram showing the placement and outcome of penalties taken by a selected footballer.

The clutter that can occur between background presentation and the foreground data representation makes it a real challenge to efficiently establish a sense of visual hierarchy. The brain and the eyes otherwise have to work especially hard to draw any insight.

What we are trying to establish is a clear sense of the most important signals brought to the foreground and the less important contextual or decorative elements pushed into the background.

We saw this effect successfully demonstrated by our proposed solution for the Olympics demonstration in the previous chapter. Here, the main focus surrounding the narrative of China and Germany's transition over time was achieved by promoting their series of values strongly into the foreground through color. The rest of the value series for the remaining countries, as well as the chart apparatus, were relegated subtly into background but were still visible and available for reference.

We see a similar effect demonstrated by this following image taken from a typical dashboard display. By their very nature, dashboards are deployed in situations whereby the efficiency and accuracy of detecting key message as signals is a key aim:

	Monthly £	Average	52 Week Sales	Best
Team A		£55.5		73
Team B		£50.0		70
Team C		£51.3		66
Team D		£53.2		73
Team E		£38.8		75
Team F		£56.0		72
Team G		£48.0		70

In this example, we see a limited, rather monochromatic color scheme applied across all properties—values, charts, labels, and titles. Through deploying this soft palette, it enables the key signals to jump in to the foreground as the most important visuals: the red indicators (alerting a need for further investigation), the blue headline bars (best performance), and the very subtle markers on the sparklines to represent the highest (blue) and lowest (orange) weekly levels.

When it comes to learning about the potential of color to create a sense of hierarchy, we can take inspiration from the effective deployment of color witnessed in other contexts. We can see examples from the best designs in advertising, website, product, and video games where creating intuitive, hierarchical displays are often vital components of their purpose and experience.

When it comes to judging background colors, there is no definitive set of rules about whether light (typically white) or dark (typically black) colors are better or worse. It is always a contextual judgment based on the intended style of the project as well as the palette of colors from which you intend to represent data. It is essentially a judgment about the legibility of contrast between foreground and background chart properties.

As a general piece of advice, try not to use strong, highly saturated colors when covering large areas. Don't force the eye to have to constantly contend with and process dominant colors. Instead, give yourself the option of using strong colors to highlight and draw attention to the data layer.

Another important property to take notice of, in the relationship between foreground and background, is the careful deployment of chart apparatus, such as the axes, gridlines, tick marks, borders, titles—any chart property you may use to frame and reference your data.

Don't be afraid to remove or dampen the visible presence of such elements, particularly as the defaults in many tools are set to black. We are automatically tempted to make things darker, bolder, more prominent, more imprisoned. Where possible, minimize, dampen, or even remove some of these chart properties because we want to let the data stand out and facilitate our "seeing" of its qualities.

This extends to elements like titles. The following are two contrasting title designs for a visualization project that was undertaken about the history of Olympic speed. The first title shows a very rich and colorful image comprising a mosaic of all the posters down the years:

When this version was incorporated on to the main design, it was immediately clear that it was too visually prominent, drawing too much attention away from the main data display. By contrast, the second version was much subtler and worked far better as a cohesive part of the final display:

THE PURSUIT OF FASTER
Visualising the evolution of Olympic speed

Image from "Pursuit of Faster" (`http://www.visualisingdata.com/index.php/2012/07/new-visualization-design-project-the-pursuit-of-faster/`), by Andy Kirk and Andrew Witherley

There are many deeper and more specific aspects of color theory around the contrast or relationship between two colors. For example, typically it can be seen that blue on black is hard for many to discriminate, as is yellow on white. There are also issues to consider about the unexpected by-product of illusions being created between different arrangements of colors and shades. Color theory is a huge field and we can only reasonably scratch the surface in this book.

To conform to design requirements

The final factor concerning color involves the necessity to incorporate an organization's visual identity and conforming to predefined color palettes. Wherever possible, you would always seek to avoid the restrictions to color choice, but often this will simply not be possible. Just imagine some of the major corporations in the world and their brand identities and you'll immediately be able to envision the definitive color palettes.

The use of predetermined color schemes in visualization is to be expected, especially because it helps maintain consistency and recognition of brand. For a designer, it can be a hindrance and so it reemphasizes the importance of identifying this requirement in your early part of the methodology.

Here is an example from the Guardian newspaper. This bubble hierarchy diagram shows the breakdown of UK Government spending by department. The image contains a wide range of colors but they hold no quantitative or categorical meaning. Aside from helping to distinguish the different clusters, they perform a largely decorative function that makes the piece more attractive to engage with and help reinforce the organization's visual identity, which is typically a very colorful spectrum:

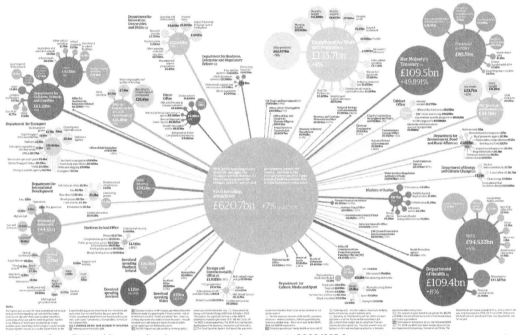

Image from "UK public spending by government department, 2008/09" (http://www.guardian.co.uk/news/datablog/2010/may/17/uk-public-spending-departments-money-cuts), by Michael Robinson and Jenny Ridley for the Guardian

Many organizations such as the Guardian and also the New York Times have developed such a strong visual identity from their respective works, consistently observing defined color palettes, that you can now immediately identify their work from the style this color usage perpetrates.

Creating interactivity

At its best, a static visualization is like a powerful photograph—a carefully conceived, arranged, and executed vision that manages to portray the sequence or motion of a story without the actual deployment of movement. In my humble opinion, delivering such an immersive experience through static design is the most elegant demonstration of data visualization.

That said, enhancements in technology over the past decade have created incredible opportunities for talented developers to construct powerful interactive visualizations. From the rapid diffusion of fast access to the Web, the development of advanced software, and programming environments, through to the immediate access to millions of live data records and the range of responsive platforms on which we can now access information, the richness of potential interactivity can lead to some incredible work.

We are currently witnessing a generation of outstanding interactive visualization projects, representing a paradigm shift in the levels of creativity, innovation, and user experience. Where once we were trapped by the limitations of a single sheet of paper, the limited real estate offered by a VDU and the slow speed of our Internet connections, now there are few, if any, genuine barriers to the potential of interactive visualizations.

Like we suggested earlier, the very best examples of interactivity manage to make the visible, invisible. That is, the functions of interactivity blend into the design so seamlessly and intuitively that the apparatus of interaction is inseparable from the data portrayal—we no longer view it as a tool wrapped around a data visualization.

Inevitably, the potential development of an interactive design requires technical capabilities. There is no way of avoiding that. Otherwise, the option to build an interactive will simply be a non-starter. Other constraints such as platform compatibility, data loading speed, and server capacity need to be factored in as well. Your ambitions may be lofty and impressive but you need to be realistic about what you can actually accomplish and this should have already been determined.

Referring back to your early thoughts about the purpose of your project, you also need to carefully consider the motivation and intention of this design. Specifically, what functional experience are you trying to create for your audience: is it an exploratory, explanatory, or maybe a combined design?

Remember, just because you can create interactivity doesn't always mean it enhances the user experience of engaging with data visualization. Don't compromise the essence of your visual communication by abandoning a static design just for the novelty of creating interactivity.

Conversely, if the complexity and variety of the data structures that you are working with make it incompatible with a static portrayal, that's exactly the situation that warrants interactive features.

If you have decided that facilitating interactivity is required for your visualization, you have many different features and functions to contemplate deploying.

The following interactive Sankey diagram is a perfect demonstration of a project that effectively integrates a host of useful interactive features that maximize the exploratory potential of the subject matter. It was developed to present a breakdown of the flow of different sources and types of fuels, from supply through transformation, and to end usage:

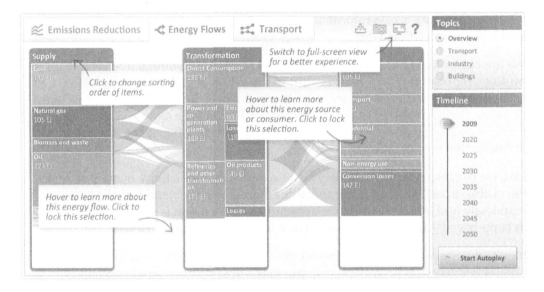

Image from "Energy Technology Perspectives 2012 Online Data" at http://www.iea.org/etp/explore © OECD/IEA 2012, developed by Raureif GmbH and Christian Behrens

Notice, through the annotated tips, the range of different actions you can trigger in order to see the data from many different perspectives. That is a key factor behind the deployment of interactivity — being able to take on multiple views of a subject matter to really understand the stories emerging.

Here is a brief outline of the variety of types of functions and features you should consider building into your interactive work:

Manipulating variables and parameters

The complexity of some data frameworks often means we are trying to find ways of showing many dimensions of stories within a single display or to facilitate different combinations of variables for exploratory visual analysis. The ability to select, filter, exclude, or modify certain variables is a valuable way of letting the user interact with different slices of the data. Furthermore, grouping and sorting options are common facilities for extracting new insights.

In the Sankey diagram example, you can isolate any of the vertical segments to see the breakdown and flow of those individual components across the entire system. You can also modify the variable of time using the slider to see changes across numerous yearly milestones.

A technique called "brushing" — highlighting a set of data marks — is another powerful way of focusing in on a subset view of our presented data, particularly with scatter plot type displays.

Adjusting the view

In contrast to manipulating variables, this is more about adjusting the user's lens or window into the subject. When we have hierarchical or high-resolution data, the ability to perform vertical exploration through the different layers of detail is an important feature. This can be particularly valuable in map-based visualizations where you may wish to pan around the landscape and zoom through different levels of magnification. You would see the benefit of this in a project such as the "Wind Map" that we saw earlier, enabling the user to dive into different parts of the country or those areas with strong winds that would be interesting to see in more detail.

An additional element of view adjustment is to create different horizontal tabs or panels of data. For example, if you wanted to show your data stories via a series of explanatory sequences. Collapsible devices such as concertinas allow for extra detail to be neatly organized hidden away from the default view and revealed when required. In the energy flow example, we also see a full-screen option that enables users to optimize the screen space occupied by the interactive.

Annotated details

We'll learn more about annotation shortly but, in interactive terms, this is about creating extra layers of data detail through interactive events such as hovering or clicking. This is particularly useful if you want to reveal actual data values or extra detail about a given category or event.

In the earlier section we discussed the degree of accuracy in interpretation and we saw an example of an interactive bubble chart. As you hovered over the bubbles, you saw a pop-up text display with the raw numbers. The availability of this type of detail, just a click or hover away from view, might give us greater creative license. By having the backup of absolute data accuracy through these values, we might give ourselves that extra confidence to choose a less precise but potentially more creative visual variable for use in our data representation display. It's almost like having a perceptual safety net.

Animation

When we have time-series based data, there is great potential for us to portray our visualization through animation, creating a shifting scene of data as it unravels a compelling data story.

The use of features such as Play, Pause, and Reset can be enhanced by offering manually controllable time sliders (seen in the earlier energy flow example) as well as chapter navigation to skip through key milestones.

The following example below, depicting the expansion of post offices across the U.S. through the years 1700 to 1900 is a perfect demonstration of the potential power of animated data presentation. While the individual frames are interesting in their own right, the real power of this portrayal comes through the emerging story of the social history of population growth and migration across the country. In the following screenshot, we see the striking moment in 1846 when the first post office on the West Coast. This is an event that would have been lost without the animated version:

Image from "Posted: Visualizing US Expansion Through Post Offices"
(http://blog.dwtkns.com/2011/posted/), created by Derek Watkins

One thing to bear in mind is that our memory capability is ill-equipped for remembering the previous scenes of an animated story. If the facilitation of comparison is important, then animation may not be the best method and something like a panel of small multiples will be more effective. The deployment of features such as trailing certain categories or the amplification of new values as they change significantly over time can also help compensate for this shortcoming.

The potential development of an interactive or animated visualization is a very exciting prospect for a talented developer and designer. However, without wishing to sound like the boring parent who doesn't let you go outside to play, the suitability and need for interactivity does need to be thoroughly reasoned and justified.

I will reemphasize the point made earlier: just because you can, doesn't mean to say you should. Interactive visualizations expand the creative opportunities but should be deployed to enhance the clarity and accessibility of data perception, not compromise it. Poorly considered clicks, sliders, filters, and menus can create unnecessary distraction and may delay access to the data and the key insights.

Annotation

Here is a quote from Amanda Cox (`http://eyeofestival.com/speaker/amanda-cox/`):

> *"The annotation layer is the most important thing we do... otherwise it's a case of here it is, you go figure it out."*

Our next layer is one that can often be neglected. However, as this quote suggests, annotating visualization is such an important features of our design. It is about taking care of your audience, recognizing who they are, what they might know already, and what they don't know.

Done well, annotation can help explain and facilitate the viewing and interpretive experience. It is the challenge of creating a layer of user assistance and user insight: how can you maximize the clarity and value of engaging with this visualization design?

As discussed in the first chapter, a key objective for effective data visualization design is the facilitation of accessibility into a subject through intuitive design. The degree of accessibility is enhanced through the effective inclusion of useful explanation across all features of your visualization solution. We shouldn't assume that readers or users are instantly and easily going to be able to navigate their way around our designs and so we need to carefully consider the best ways to assist them; explained as follows:

- **Titles**: A compelling title can help to attract an audience and articulate the focus of the visualization subject matter. Sometimes, especially in explanatory visualizations, you can look to exploit this prominent space to tell readers about a key insight or headline. However, make sure it is an accurate reflection of the content of the visualization otherwise it will be misleading.

- **Introductions**: These are really important instructive elements to explain t he project's background and context, describing the background motivation and what your intentions are in terms of how it should be used.

- **User guides**: While intuitive accessibility is stated as an overall goal, many projects often warrant further explanation, particularly with interactive pieces and those that have inherently complex subjects or frameworks.

  In this next project, titled *Political Moneyball* and created by the Wall Street Journal, we see a demonstration of exceptional care for the audience's understanding of how to optimize this visualization's use. Not only does it include thorough written annotation and labeling to help users understand all the features of this incredibly immersive tool, but there is also a video tutorial to offer that extra degree of support. The designers of this piece astutely recognize the potential depth and interpretive complexity of the subject matter and I imagine also want to do justice to their efforts to bring this deep subject to fruition.

Search in this column includes parties, industries, companies and individual donors.

This gray panel displays the name of the committee, the total amount raised. (Hint: click 'List view' for a ranking)

Use these buttons to go back to the last screen, reset to the beginning, share your findings or zoom in and out.

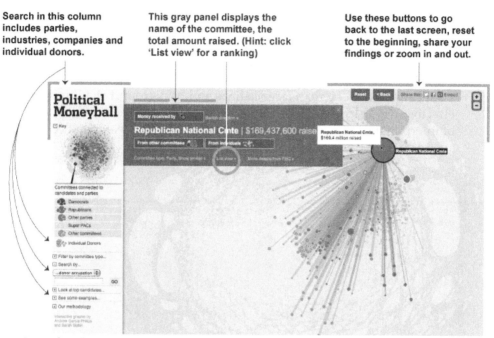

Image from "Political Moneyball" (http://graphics.wsj.com/political-moneyball/#), created by Andrew Garcia Phillips and Sarah Slobin of the Wall Street Journal

- **Labels**: In the interactive section, we discussed the potential of labels to reveal extra details about data values. As we see in the previous project, labeling is an incredibly simple but useful device to help explain matters. Often, these are hidden and interactively revealed through selection or by hovering.

- **Captions and narrative**: In addition to the potential use of the title to offer a key headline, sometimes you may wish to surface important insights and findings to help fast-track the reader's interpretation process. You might draw out the good and the bad or maybe the expected or unexpected. You should also consider the potential value, in certain projects, of the "what next?" question—what should the user do with this information? what actions need to be taken?

- **Visual annotation**: Annotation goes beyond just written explanations and we should consider how to use chart or graphic devices to help draw out important insights visually. Simple options include features such as gridlines, axes labels, and tick marks. In *Chapter 3, Demonstrating Editorial Focus and Learning About Your Data*, we saw an example of effective visual annotation. Here, reference lines and background shading is used effectively to help the reader achieve distinction between different tiers of interpretation, as you explore the relationship between what countries spend on education and the military.

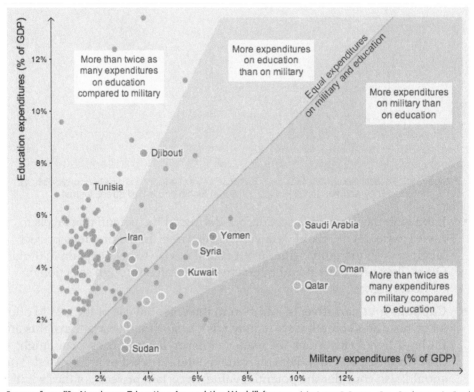

Image from "In Numbers: Education Around the World" (http://visualdata.dw.de/specials/bildung/en/index.html), created by Gregor Aisch for Deutsche Welle

- **Legends and keys**: Always explain the use of color schemes or the varying size of shapes in terms of their categorical or quantitative representation.

- **Units**: You should include details of the units of values being displayed to ensure you don't create ambiguities and potential misinterpretation. As with many of these annotated features, this is an obvious requirement, something we've had drilled into us since our school days, but you'd be surprised how often they can be left out.

- **Data sources**: It is vital to include detailed references about from where you have accessed your data or any other sourced element (such as imagery). Where you have chance to offer a more detailed narrative, you may wish to explain what treatment you have applied to the data in terms of its quality or analytical transformation.

- **Attribution**: Don't forget to acknowledge those who have either contributed directly, influenced the construction of the design, or those people whose work has acted as a source inspiration.

The final thing to mention about annotation is that this is likely to be the first time we have to consider our typography selections. There are, of course, plenty of established guides and sources of literature to help influence your choice of fonts for all pieces of written annotation. However, this is another aspect of design that you will be able to ultimately judge best using your own design instinct. Many designers have their favorites and like to maintain this identity but also many projects may be required to observe certain visual identity rules like we outlined in the color section.

Arrangement

You have established how you are going to represent your data, you've identified your visual identity through color, the choices around static or interactive design have been rationalized, and you have identified the range of annotation requirements.

For our final layer, we need to consider how to arrange our design in terms of the layout, placement, and organization of all visible elements. How can we piece everything together most effectively?

As we've just discussed in relation to annotation, our intention with the arrangement and architecture of our design is to deliver as intuitive an experience as possible. The level of intuitiveness and smooth access into the subject matter is strongly influenced by the logic and implied meaning behind the arrangement of our chart elements, the interactive features, and annotation devices.

The key overall aim is to reduce the amount of work the eye has to undertake to navigate around the design and to decipher the sequence and hierarchy of the display. For the brain, once again, we're looking to minimize the amount of thinking and "working out" that goes on. We therefore need to carefully consider the choices we make around the size, positioning, grouping, and sorting of all that we show. As with all visualization design layers, we need to be able to justify the decisions we make about every visible property presented.

Here is a simple, but effective, demonstration of the careful consideration of arrangement. It is just one example out of many we could refer to from the projects shown in this book.

Observe the positioning of the chapter navigation slider across the top, the size of the space afforded to the main map display, the narrative found on the right-hand side, the proximity of the legend to the data, and the location of the pan and zoom device—all these decisions are very deliberate and designed to maximize the logic and meaning behind the layout of this project's data, its interactive features and annotated elements.

Image from "The Growth of Newspapers Across the U.S.: 1690-2011" (`http://www.stanford.edu/group/ruralwest/cgi-bin/drupal/visualizations/us_newspapers`), created by Rural West Initiative, Bill Lane Center for the American West, Stanford University.

On the matter of arrangement, it is important to mention an important paper produced by Edward Segal and Jeff Heer of the Stanford Vis group and titled *Narrative Visualization: Telling Stories with Data* (`http://vis.stanford.edu/papers/narrative`).

As the title suggests, this article provides an excellent outline of the different design strategies for arranging and structuring the layout of your visualizations that will help maximize the potential telling of stories through data.

Summary

In this chapter, we have assessed the major design decisions we need to take across the key layers of a data visualization's anatomy. We have developed a better understanding of the choices we need to make and a sense of how to rationalize them depending on the context of our project.

We have worked through the challenge of selecting the appropriate data representation solution and then pulled apart the different aspects of data presentation in order to create a cohesive design concept.

Some might consider this to be an overly systematic or scientific way to tackle a task that has significant elements of creativity. However, this approach is consistent with the central theme of this book, to equip you with effective tactics and strategies to move gracefully through this potentially tricky design concepting process.

Remember that even the masters of the subject struggle to avoid making mistakes and ending up in creative blind alleys. A worthwhile design is rarely arrived at instantly and without the need for iterations and rethinks.

A key message is not to put yourself under too much pressure to get things right, straightaway and every time. Hopefully, following this framework of decisions and options will give you as good a chance as possible to get to your solution quicker and more efficiently through practice and experience.

In the next chapter, we will look closely at the taxonomy of different visualization methods and the various common chart types techniques that fall under each category. This will help cement your understanding of the data representation layer, undoubtedly the most important design feature to get right, and provide you with a convenient gallery of options to consider for your own projects.

5

Taxonomy of Data Visualization Methods

In the previous chapter, we learned about the anatomy of a data visualization. This provided a framework to recognize and consider all the different design decisions you're faced with as well as guidance for how you might rationalize the choices you make.

In this chapter, we specifically revisit the data representation layer. This was described as the most critical layer of the visualization design task. It is also probably the most difficult to master.

In this chapter, we'll look at a taxonomy of data visualization methods as defined by the primary communication purpose. Within this taxonomy we will see an organized collection of some of the most common chart types and graphical methods being used today.

By exploring this array of chart types you will get a better understanding of the relationship between the stories you are trying to portray, the physicality of your data, and the visual variables through which we can represent them. It will provide you with a catalog of reference, offering ideas, and inspiration for when you face this stage in the design of your own projects.

The gallery of solutions presented does not intend or pretend to cover every subtle variation of chart or graph design. A creative field like data visualization simply does not lend itself to finite classification.

However, it should help you feel better equipped to more efficiently determine the most suitable representation solution for your specific problem context.

Data visualization methods

The common definition for taxonomy comes from the biological sciences and refers to the organization into groups of members that share similar characteristics. In this case, the members are chart types and the shared characteristic is based on the primary data portrayal function.

Selecting the appropriate visualization method will be influenced by the definition work you undertook earlier in the methodology to clarify the intention of your visualization communication.

It is about starting the journey towards identifying the most suitable way to answer your main data questions: how are you going to show, what it is you want to say.

Here is an outline of the primary communication purpose of each method classification:

Method classification	Communication purpose
Comparing categories	To facilitate comparisons between the relative and absolute sizes of categorical values. The classic example would be the bar chart.
Assessing hierarchies and part-to-whole relationships	To provide a breakdown of categorical values in their relationship to a population of values or as constituent elements of hierarchical structures. The example here would be the pie chart.
Showing changes over time	To exploit temporal data and show the changing trends and patterns of values over a continuous timeframe. A typical example is the line chart.
Plotting connections and relationships	To assess the associations, distributions, and patterns that exists between multivariate datasets. This collection of solutions reflects some of the most complex visual solutions and usually focuses on facilitating exploratory analysis. A common example would be the scatter plot.
Mapping geo-spatial data	To plot and present datasets with geo-spatial properties via the many different mapping frameworks. A popular approach would be the choropleth map.

Once we have selected the appropriate method, we then start to work through the other key ingredients of the data representation selection process, as outlined in *Chapter 4, Conceiving and Reasoning Visualization Design Options.*

Weighing up the following factors helps us to narrow down the variety of options within each method classification to find the most suitable choice of specific chart type or graphical method:

- Does it accommodate the physical properties of your data?

- Does it facilitate the desired degree of accuracy?

- Is it potentially capable of conveying a certain metaphorical and design consistency with our subject matter?

Choosing the appropriate chart type

Attempting to organize chart types based on their primary portrayal method is not new (see `http://queue.acm.org/detail.cfm?id=1805128` and `http://www.visualizing.org/stories/taxonomy-data-visualization`). The classifications presented in this chapter reveal a personal view—informed by knowledge, experience, and instinct—of a logical way to organize thinking about the relationship between data variables, visual variables, and chart frameworks.

The examples presented are intended to cover the most typical and contemporary approaches being used today. It is an arbitrary statement, but you should find that on 95 percent of occasions one or several of the chart types shown will cover your requirements. The remaining 5 percent will probably require custom-built solutions for very specific data shapes and contexts.

Note that many of the chart types presented hold numerous presentational characteristics and could belong to more than just one classification of method. For example, an area chart shows changes over time and enables comparison between categories. As the chart types have been organized based on their primary method, the prominence of the time-series nature of this example would lend itself more towards the "showing changes over time" method category.

As you go through the chapter, you will see the following information:

- The popular and alternative names used to describe each chart.

- The type and quantity of typical data variables you would normally use with each type of chart. On most occasions *any* categorical or quantitative variable is suitable though more specific variable types (nominal, ordinal, ratio-, interval-scale) are proposed where applicable.

- The visual variables that have been used to represent data (optional variables are italicized) in each chart.

- A brief description of each type's functional purpose and application.

- An example to illustrate what each chart looks like. Many of these have been seen elsewhere in this book so it will hopefully cement your understanding.

Remember, you may often need a combination of different visualization methods and multiple chart types blended together to forge a multidimensional story.

Comparing categories

The following examples present chart types that facilitate the comparison of categorical values.

Dot plot

Data variables: 2 x categorical, 1 x quantitative.

Visual variables: Position, color-hue, symbol.

Description: A dot plot compares categorical variables by representing quantitative values with a single mark, such as a dot or symbol. The use of sorting helps you to clearly see the range and distribution of values. You can also combine multiple categorical value series on to the same chart distinguishing them using color or variation in symbol. Beyond two series things do start to get somewhat cluttered and hard to read.

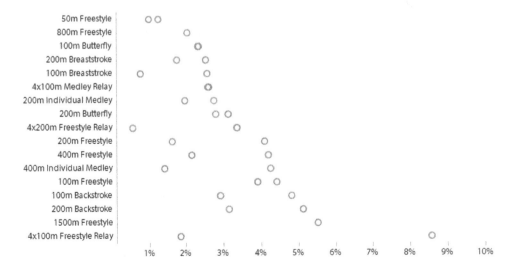

Bar chart (or column chart)

Data variables: 1 x categorical, 1 x quantitative-ratio.

Visual variables: Length/height, color-hue.

Description: Bar charts convey data through the length or height of a bar, allowing us to draw accurate comparisons between categories for both relative and absolute values. When using length as the visual variable to represent a quantitative value it is important to show the full extent of this property so always start the bar from the zero point on the axis. The use of color can help draw attention to the values of specific categories in accordance with your narrative, as shown in the following screenshot:

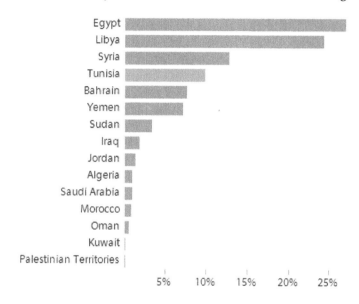

Floating bar (or Gantt chart)

Data variables: 1 x categorical-nominal, 2 x quantitative.

Visual variables: Position, length.

Description: A floating bar chart—sometimes labeled a Gantt chart because of similarities in appearance—helps to show the range of quantitative values. It presents a bar stretching from the lowest to the highest values (therefore the starting position is not the zero point). Using such charts enables you to identify the diversity of measurements within a category and view overlaps and outliers across all categories. A Gantt chart is shown in the following screenshot:

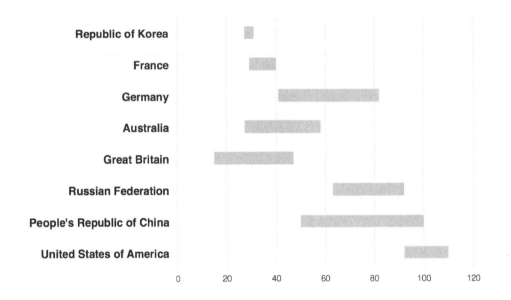

Pixelated bar chart

Data variables: Multiple x categorical, 1 x quantitative.

Visual variables: Height, color-hue, symbol.

Description: The proposed name of "pixelated bar chart" is more an intuitive description than an established type. These charts provide a dual layer of resolution: a global view of a bar chart (showing aggregate totals) and a local view of the detail that sits beneath the aggregates (demonstrated by the pixels shown within each bar). Typically, these charts are interactive and offer an ability to hover over or click on the constituent pixels/symbols to learn about the stories at this more detailed resolution.

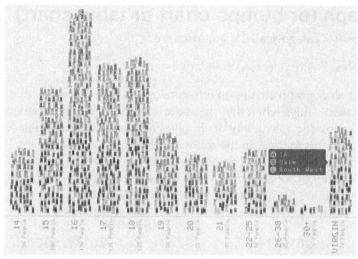

Image from "The Sexperience 1000" (`http://sexperienceuk.channel4.com/the-sexperience-1000`), created by Channel 4/Mint Digital

Histogram

Data variables: 1 x quantitative-interval, 1 x quantitative-ratio.

Visual variables: Height, width.

Description: Histograms are often mistaken for bar charts but there are important differences. Histograms show distribution through the frequency of quantitative values (y axis) against defined intervals of quantitative values(x axis). By contrast, bar charts facilitate comparison of categorical values. One of the distinguishing features of a histogram is the lack of gaps between the bars, as shown in the following image:

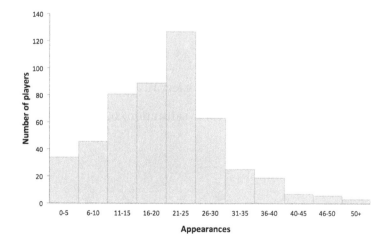

Slopegraph (or bumps chart or table chart)

Data variables: 1 x categorical, 2 x quantitative.

Visual variables: Position, connection, color-hue.

Description: A slopegraph creates an effective option for comparing two (or more) sets of quantitative values when they are associated with the same categorical value. They especially provide a neat way of showing a before and after view or comparison of two different points in time. In the following example, we see the total points won for teams in the English Premier League across two comparable seasons. The layout creates a combined view of rank and absolute value based on position on the vertical axis, with a link joining the associated values to highlight the transitional change. Color can be used to further emphasize upward or downward changes:

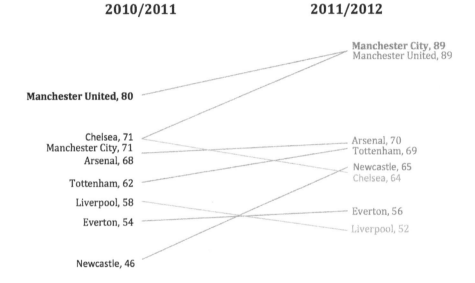

Radial chart

Data variables: Multiple x categorical, 1 x categorical-ordinal.

Visual variables: Position, color-hue, color-saturation/lightness, texture.

Description: A radial chart displays data around a concentric, circular layout. The example shown in the following image shows the status status across a number of different categorical measures relating to gay rights for each state in the U.S., arranged to indicate approximate geographical relationships. A slight visual shortcoming associated with a radial chart is the fractionally distorted prominence of the segments on the outside rings which end up being larger (due to arc length) than those on the inside. Often radial charts are used for showing data over time but this only works when the sequence is continuous (such as a 24 hour clock).

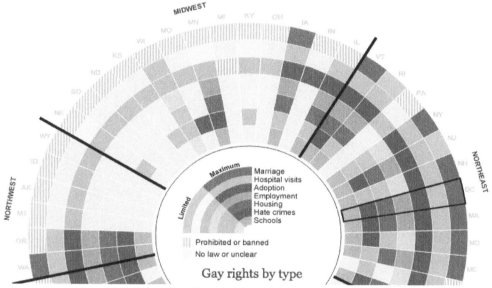

Image from "Gay rights in the US, state by state" (http://www.guardian.co.uk/world/interactive/2012/may/08/gay-rights-united-states), created by Guardian in America Interactive

Glyph chart

Data variables: Multiple x categorical, multiple x quantitative.

Visual variables: Shape, size, position, color-hue.

Description: A glyph chart is based on a shape (in the following example, a flower) being the main artifact of representation. The physical properties of the shape (through a feature such as a petal) represent different categorical variables; they are sized according to the associated quantitative value and distinguished through color. While absolute magnitude judgments are not easily achieved nor intended, the hierarchy of the data (big, medium, and small values) is possible to interpret and the typical deployment of interactivity enables further exploration.

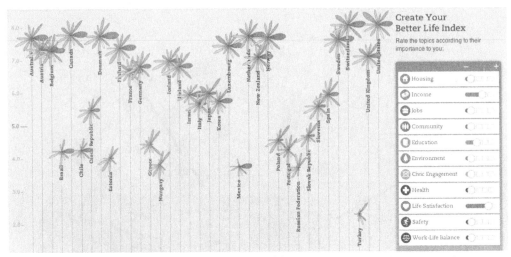

Image from "OECD Better Life Index" (http://oecdbetterlifeindex.org), created by Moritz Stefaner (htpp://moritz.stefaner.eu) in collaboration with Raureif GmbH (http://raureif.net)

Sankey diagram

Data variables: Multiple x categorical, multiple x quantitative.

Visual variables: Height, position, link, width, color-hue.

Description: Sankey diagrams are used to convey the idea of flow. They portray constituent quantities of a series of associated categorical values, across a number of "stages", with the ongoing associations represented by connecting bands. The width of these links indicates the proportional flow from one stage to another. They are useful for showing situations where elements transform and divide over key events, as shown here displaying the breakdown of different fuels, how they are transformed and then ultimately used.

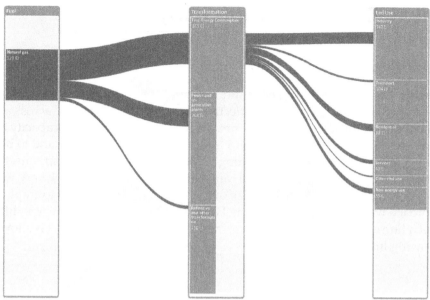

Image from "Energy Technology Perspectives 2012 Online Data" at http://www.iea.org/etp/explore © OECD/IEA 2012, developed by Raureif GmbH and Christian Behrens

Area size chart

Data variables: 1 x categorical, 1 x quantitative-ratio.

Visual variables: Area, color-hue.

Description: This type of chart doesn't have an obvious name, so Area size chart is a best attempt! It is a very simple visual device that deploys the visual variable of area—normally a rectangle or circle—to compare two (or maybe several) values. Normally these values will vary in size quite dramatically to convey a certain shock at the disparity. The subject matter may relate to a part-of-a-whole comparison (portion judgment) but more typically involves separate, independent categories (comparative judgment).

A Tale of Two Leagues: Comparing Transfer Spend (Summer 2012)

English Premier League
£554M

Scottish Premier League
£22M

Small multiples (or trellis chart)

Data variables: Multiple x categorical, multiple x quantitative.

Visual variables: Position, any visual variable.

Description: Small multiples are not really a separate chart type but an arrangement approach that facilitates efficient and effective comparisons to be made across a multipanel display of small chart elements. These displays exploit the capacity of our visual system to rapidly scan across a trellis of small similar charts and to be capable of easily and immediately spotting patterns. These are particularly useful for comparing categories that have a broad range of values. They also work very well for showing snapshots of events that change over time. One of the earliest examples of this approach was *The Horse in Motion* by Eadweard Muybridge to show the different stages of a horse's movement over a time frame-by-frame. A trellis chart is shown in the following image:

Word cloud

Data variables: 1 x categorical, 1 x quantitative-ratio.

Visual variables: Size.

Description: Word clouds depict the frequency of words used in a given set of text. The font size indicates the quantity of each word's usage. Color is often just used as decoration (which you'll notice actually distorts the visual prominence). While it's fair to say they are becoming something of a ubiquitous visual commodity, they can be useful for exploring datasets for the first time in order to identify key terms being used. If you feel compelled to use word clouds, the best advice is to ensure the underlying text being used is carefully prepared in advance to reduce the noise. A word cloud is shown here:

Assessing hierarchies and part-to-whole relationships

The following examples present chart types that help us to assess hierarchical and part-to-whole relationships.

Pie chart

Data variables: 1 x categorical, 1 x quantitative-ratio.

Visual variables: Angle, area, color-hue.

Description: Pie charts are probably the most contentious chart type and attract much negative sentiment. While we know it is harder to accurately interpret angles and judge the area of segments compared to other visual variables, the negativity is arguably more a reflection of their relentless misuse. The inclusion of too many categories and colors, 3D decoration, and poorly executed arrangement are often to blame for this. Usually, a simple bar chart will suffice to demonstrate the part-to-whole relationship. However, if you are determined to use a pie chart, always start the first slice from the vertical position (to establish a sense of baseline), minimize the number of categories being displayed (ideally maximum of three), and arrange the segments as logically as possible. Variations include the donut chart, which is essentially the same chart but with the center removed (to accommodate labels or nested donut charts).

League Within a League: Total Transfer Spend, Premier League 2012

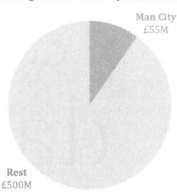

Stacked bar chart (or stacked column chart)

Data variables: 2 x categorical, 1 x quantitative-ratio.

Visual variables: Length, color-hue, position, color-saturation/lightness.

Description: Stacked bars are fairly self-explanatory. They can be based on the stacks of absolute values or standardized to show part of a whole breakdown, as in following example. Colors and position differentiate the value categories. Where the categorical values are ordinal in nature, it helps to sequence the values logically, for example when you have sentiment data such as the Likert scale of disagree (reds) through to agree (blues). This sequencing helps draw out the contrasting composition of the sentiment from all categories. The only drawback of a stacked chart is the difficulty in being able to accurate read bar lengths, as there is no common baseline.

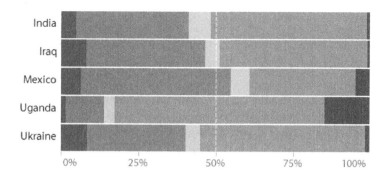

Square pie (or unit chart or waffle chart)

Data variables: 1 x categorical, 1 x quantitative-ratio.

Visual variables: Position, color-hue, symbol.

Description: There are several titles for this type of chart but the common technique involves a grid of units (may be squares or symbols) to represent parts of a whole. This may be for a percentage comparison (square pie) or an absolute quantity (unit chart, waffle chart). The use of color and symbol establishes the visual composition of the categorical and quantitative values. An example is shown here:

Champions vs. Promoted Teams: Total Transfer Spend, Premier League 2012

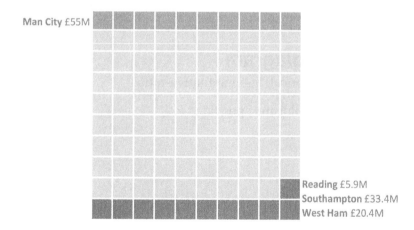

Tree map

Data variables: Multiple x categorical-nominal, 1 x quantitative-ratio.

Visual variables: Area, position, color-hue, color-saturation/lightness.

Description: Tree maps take the concept of a whole population and divide up portions of rectangular spaces within to represent organized, clustered constituent units sized according to their relative value. As well as arrangement, various properties of color are typically used to provide additional layers of quantitative or categorical insight. Here is an example:

Image from "Newsmap" (http://newsmap.jp/), created by Marcos Weskamp

Circle packing diagram

Data variables: 2 x categorical, 1 x quantitative-ratio.

Visual variables: Area, color-hue, position.

Description: As the title suggests, this type of chart seeks to pack together constituent circles into an overall circular layout that represents the whole. Each individual circle represents a different category and is sized according to the associated quantitative value. Other visual variables, such as color and position, are often incorporated to enhance the layers of meaning of the display. Note that you can't tessellate circles and so the combined view never creates a perfect fit (there are always gaps). The algorithms used to form the arrangement will often utilize certain overlapping properties to maintain the accuracy of the respective part-to-whole area sizes.

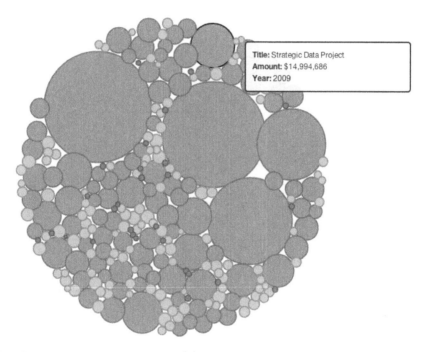

Image from "Gates Foundation Educational Spending" (http://vallandingham.me/vis/gates/), created by Jim Vallandingham

Bubble hierarchy

Data variables: Multiple x categorical, 1 x quantitative-ratio.

Visual variables: Area, position, color-hue.

Description: This technique is used to portray organization and structure through a hierarchical display. In the following example, we see the use of circles to represent the constituent departments, sized according to their quantitative value and colored to visually distinguish the different departments.

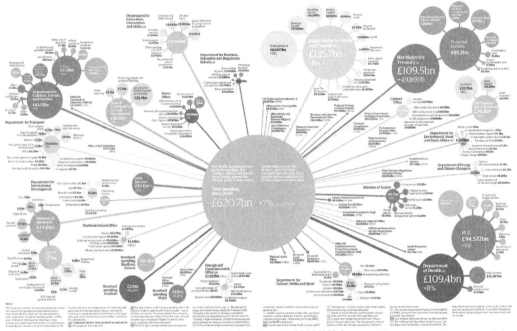

Image from "UK public spending by government department, 2008/09" (http://www.guardian.co.uk/news/datablog/2010/may/17/uk-public-spending-departments-money-cuts), created by Michael Robinson and Jenny Ridley for the Guardian

Tree hierarchy

Data variables: 2 x categorical, 1 x quantitative-ratio.

Visual variables: Angle/area, position, color-hue.

Description: Similar to the bubble hierarchy, this technique presents the organization and structure of data through a hierarchical tree network. In the following example, portraying the structure of a book, the effect is quite abstract but every visual property is serving the purpose of representing just the data - the quantitiative properties and hierarchical arrangement of all the book's elements:

Image from "Literary Organism" (`http://itsbeenreal.co.uk/index.php?/wwwords/literary-organism/`), created by Stefanie Posavec

Showing changes over time

The following examples show alternative ways of graphically showing changes over time:

Line chart

Data variables: 1 x quantitative-interval, 1 x quantitative-ratio, 1 x categorical.

Visual variables: Position, slope, color-hue.

Description: Line charts are something we should all be familiar with. They are used to compare a continuous quantitative variable on the x axis and the size of values on the y axis. The vertical points are joined up using lines to show the shifting trajectory through the resulting slopes. Line charts can help unlock powerful stories of the relative or (maybe) related transition of categorical values. Unlike bar charts, the y axis doesn't need to start from zero because we are looking at the relative pattern of the data journey.

An example is shown in the following image:

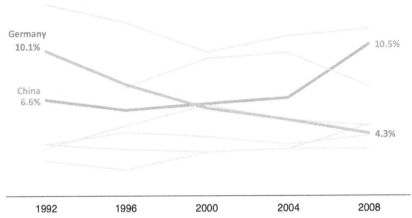

Sparklines

Data variables: 1 x quantitative-interval, 1 x quantitative-ratio.

Visual variables: Position, slope.

Description: Sparklines aren't necessarily a variation on the line chart, rather, a clever use of them. They were conceived by Edward Tufte and are described as "intense, word-sized graphics". They take advantage of our visual perception capabilities to discriminate changes even at such a low resolution in terms of size. They facilitate opportunities to construct particularly dense visual displays of data in small space and so are particularly applicable for use on dashboards. An example is shown here:

	Monthly £	Average	52 Week Sales	Best
Team A		£55.5		73
Team B		£50.0		70
Team C		£51.3		66
Team D		£53.2		73
Team E		£38.8		75
Team F		£56.0		72
Team G		£48.0		70

Area chart

Data variables: 1 x quantitative-interval, 1 x categorical, 1 x quantitative-ratio.

Visual variables: Height, slope, area, color-hue.

Description: As you can see in the following example, a number of visual properties are involved in area charts. The vertical position and connecting slope of the horizon (like a line chart) shows the progression of the values over time and the color area underneath the chart helps to emphasize these changes. Unlike a standard line chart, an area chart should have the y axis starting at zero to ensure the area judgment is being interpreted accurately.

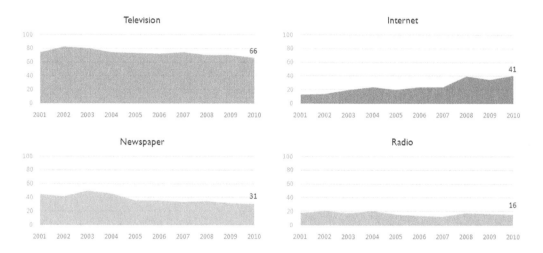

Horizon chart

Data variables: 1 x quantitative-interval, 1 x categorical, 2 x quantitative-ratio.

Visual variables: Height, slope, area, color-hue, color-saturation/lightness.

Description: This is a variation on the area chart, modified to include (and cope with) both positive and negative values. Rather than presenting negative values beneath the x axis, the negative area is mirrored on to the positive side and then colored differently to indicate its negative polarity. The result is a chart that occupies a single row of space, which helps to accommodate multiple stories onto a single display and facilitates comparison to pick out local and global patterns of change over time.

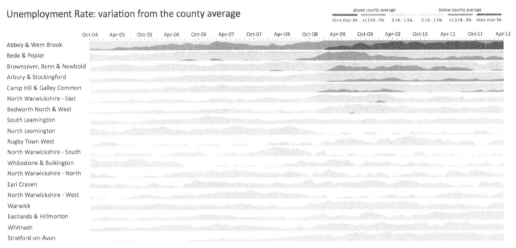

Image from "Unemployment Rate: variation from the county average" (http://warksobservatory. files.wordpress.com/2012/07/unemployment-horizon-chart.pdf), created by Spencer Payne/Warwickshire Observatory

Stacked area chart

Data variables: 1 x quantitative-interval, 1 x categorical, 1 x quantitative-ratio.

Visual variables: Height, area, color-hue.

Description: A stacked area chart provides a compositional view of categories to show their changes over time. As the title suggests, these are based on stacks of area charts differentiated by color and present either absolute aggregates or percentage aggregates. Note that the quantitative values are represented by the height (derived from top and bottom positions) of the area stacks at any given point. Sometimes the resulting shapes of the middle sections can be slightly misleading and misinterpreted due to the lack of a common baseline position.

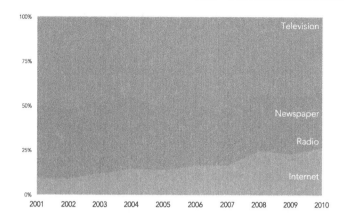

Stream graph

Data variables: 1 x quantitative-interval, 1 x categorical, 1 x quantitative-ratio.

Visual variables: Height, area, color-hue.

Description: The stream graph operates in a similar fashion to a stacked area chart, allowing multiple values series to be layered as streams of area with quantitative values expressed through the height of the individual stream at any given time. It has no baseline x axis and so there is no concept of negative or positive values, purely aggregates. Its functional purpose is really to highlight peaks and troughs—it has a particularly organic feel and is suited to displays intended to show "ebb and flow" stories. Many stream graphs will offer interactivity to allow you to explore and isolate individual layers. An example is shown in the following screenshot:

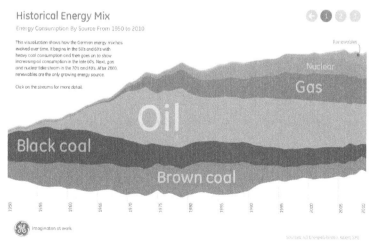

Image from "German Energy Landscape" (http://visualization.geblogs.com/ visualization/germanenergy), created by Gregor Aisch

Candlestick chart (or box and whiskers plot, OHLC chart)

Data variables: 1 x quantitative-interval, 4 x quantitative-ratio.

Visual variables: Position, height, color-hue.

Description: The candlestick chart is commonly used in financial contexts to reveal the key statistics about a stock market for a given timeframe (often daily). In the following example, we see stock market changes by day based on the OHLC measures—opening, highest, lowest, and closing prices. The height of the central bar indicates the change from the opening to closing price and the color tells us if this is an increase or decrease. This is a chart that clearly requires a certain amount of experiential learning in order to read it efficiently. However, one you've achieved this you will see how extremely dense and powerful these displays are. They are similar in concept to the "box and whiskers plot", which focus on the statistical distribution of a set of values (showing upper and lower quartiles as well as the median).

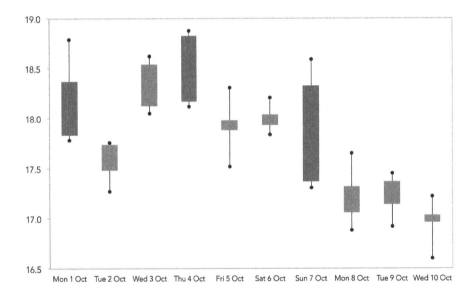

Barcode chart

Data variables: 1 x quantitative-interval, 3 x categorical.

Visual variables: Position, symbol, color-hue.

Description: These are very compact displays that depict a sequence of events or milestones over the course of time using a combination of symbols and color. In the following example, we see the key events during two football matches. Demonstrating similar qualities to those of a sparkline, barcode charts (named because they look like barcodes, funnily enough) convey a significant amount of data packed into a small space. Once again, as you familiarize yourself with how to read these charts, they do unlock a terrific amount of narrative.

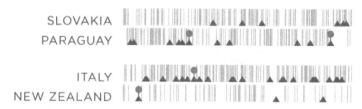

Image from "Umbro World Cup Poster" (`http://www.mikemake.com/Umbro-s-World-Cup-Poster`), created by Michael Deal

Flow map

Data variables: Multiple x quantitative-interval, 1 x categorical, 1 x quantitative-ratio.

Visual variables: Position, height/width, color-hue.

Description: Similar in many ways to the Sankey diagram, a flow map portrays the flow of a quantitative value as it is transformed over time and/or space. In this famous example, showing the march of Napoleon's army in the Russian campaign of 1812, the thickness of the main band indicates the size of the army as it moves over time and geography towards Moscow. The geographical accuracy of the plot is preserved in this chart but we don't see (or need to see) the full map detail. Notice too that the freezing temperatures are presented in the line chart below the main display, providing a further layer of the detail behind this story:

The following map shows Napoleon's famous Russian campaign:

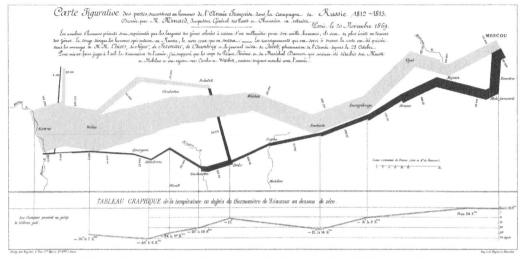

Images republished from the freely licensed media file repository Wikimedia Commons, source:
http://commons.wikimedia.org/wiki/File:Minard.png

Plotting connections and relationships

We now look at the different visualization techniques used to plot connections and relationships:

Scatter plot

Data variables: 2 x quantitative.

Visual variables: Position, color-hue.

Description: A scatter plot is a combination of two quantitative variables plotted on to the x and y axes in order to reveal patterns of correlations, clustering, and outliers. This is a very important chart type, in particular, for when we are familiarizing with and exploring a dataset. An sample scatter plot is shown in the following image:

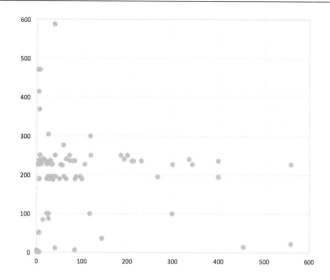

Bubble plot

Data variables: 2 x quantitative, 2 x categorical.

Visual variables: Position, area, color-hue.

Description: A bubble plot extends the potential of a scatter plot through multiple encoding of the data mark. In the following example, we see the marks becoming circles of varying size and then colored according to their categorical relationship. Often, you will see a further layer of time-based data applied to convey motion with the plot animated over time.

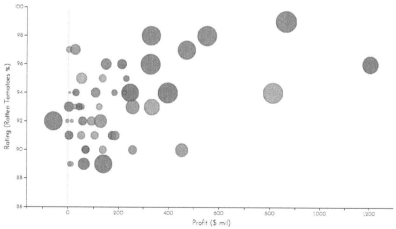

Image from "How Much Money do the Movies we Love Make?"
(http://vallandingham.me/vis/movie/), created by Jim Vallandingham

Scatter plot matrix

Data variables: 2 x quantitative, 2 x categorical.

Visual variables: Position, color-hue.

Description: Similar to the small multiples chart that we saw earlier, a scatter plot matrix takes advantage of the eye's rapid capability to spot patterns across multiple views of the same type of chart. In the following case, we have a panel of multiple combined scatter plots:

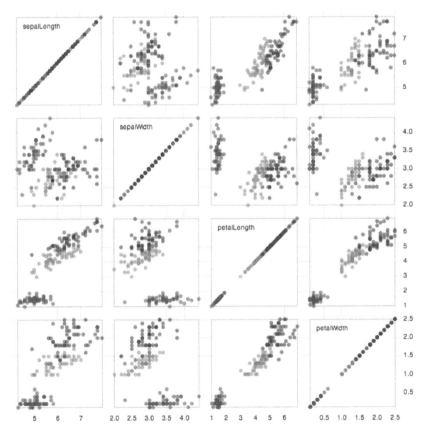

Image from "Scatterplot Matrix" (http://mbostock.github.com/d3/ex/splom.html), created by Mike Bostock

Heatmap (or matrix chart)

Data variables: Multiple x categorical, 1 x quantitative-ratio.

Visual variables: Position, color-saturation.

Description: With further similarities to small multiples, heatmaps enable us to perform rapid pattern matching to detect the order and hierarchy of different quantitative values across a matrix of categorical combinations. The use of a color scheme with decreasing saturation or increasing lightness helps create the sense of data magnitude ranking.

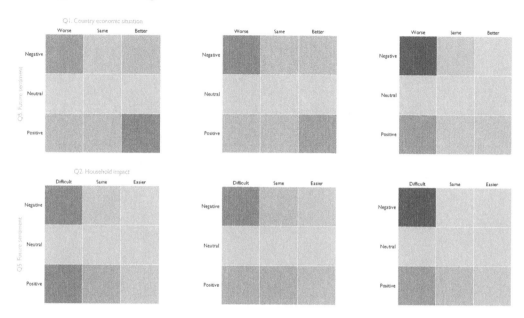

Parallel sets (or parallel coordinates)

Data variables: Multiple x categorical, multiple x quantitative-ratio.

Visual variables: Position, width, link, color-hue.

Description: Parallel sets offer a unique way of visually exploring and analyzing datasets. The technique involves plotting all your data on to a series of axes, one for each of the variables you are interested in examining. This creates pathways that show the connections between the breakdown of values contained within your data for each variable. They are useful for learning about the potential correlations and consistencies that exist in our datasets. You'll notice certain similarities with the function of Sankey diagrams.

An example of parallel sets is given in the following image:

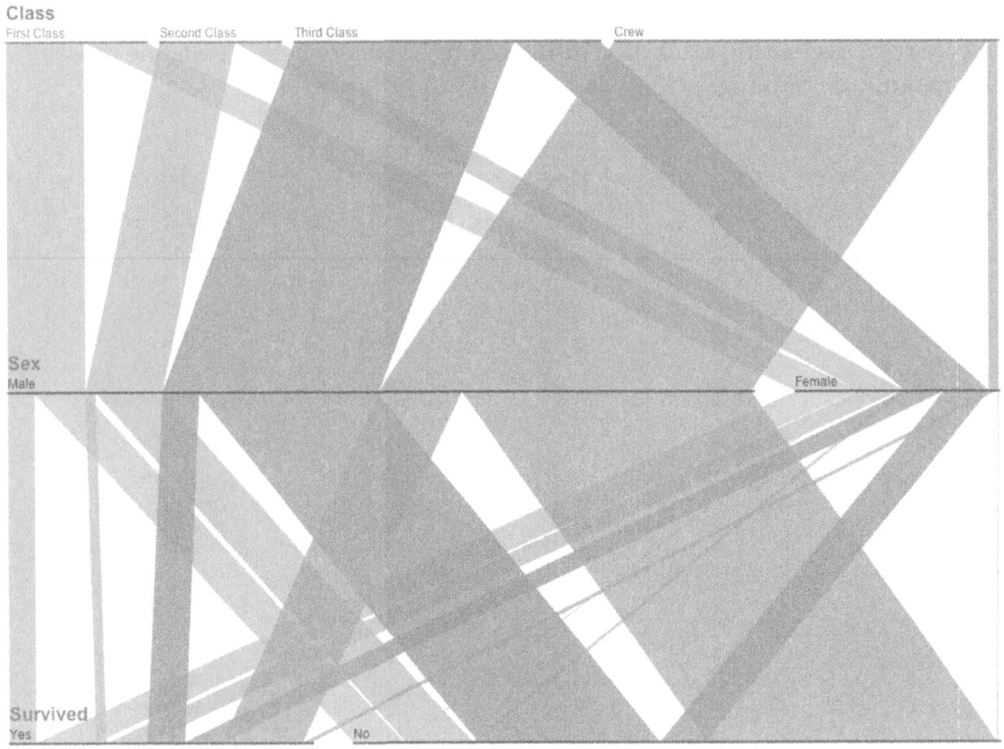

Image from "Parallel Sets" (http://eagereyes.org/parallel-sets),
created by Robert Kosara and Caroline Ziemkiewicz

Radial network (or chord diagram)

Data variables: Multiple x categorical, 2 x quantitative-ratio.

Visual variables: Position, connection, width, color-hue, color-lightness, symbol, size.

Description: A radial network or chord diagram creates a framework for comparing complex relationships between categorical values. The use of a radial layout offers the opportunity to move beyond the restrictions of an x and y axis pairing. The key explanatory property is the connections that exist between components, sometimes sized (thickness) and colored to incorporate extra layers of detail. In the following example, we see additional levels of detail represented by the encoded size of text and icons. One thing to note is that the length (and therefore prominence) of a line can slightly mislead by inferring significance (a more important relationship) when it is simply a by-product of the arrangement around the radial layout.

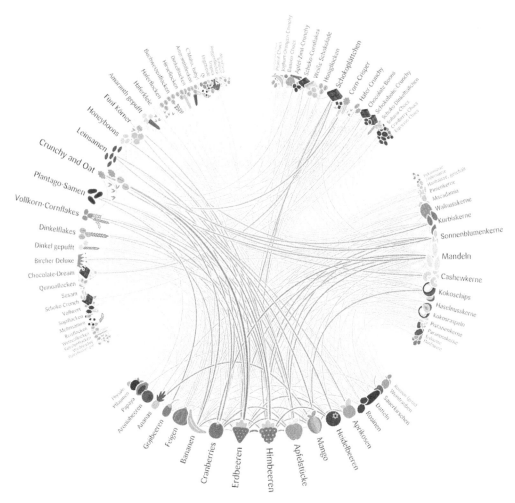

Image from "Müsli Ingredient Network" (`http://moritz.stefaner.eu/projects/musli-ingredient-network/`), created by Moritz Stefaner for mymuesli (`http://mymuesli.com`)

Network diagram (or force-directed/node-link network)

Data variables: Multiple x categorical-nominal, 1 x quantitative-ratio.

Visual variables: Position, connection, area, color-hue.

Description: At first glance, network diagrams, similar to the one shown in the following example, can look quite daunting through their visual complexity and apparent clutter (indeed, often they are described as "hairballs"). Their intention and value is to facilitate exploration of complex data frameworks based on the existence or quantifiable strength of relationships, connections, and logical organization. The typical purpose of these graphs is to enable the viewer to get a sense of patterns—picking out the elements that are of interest, observing clusters and gaps, dominant nodes and sparse connections. There are many derivatives of network diagrams with variations influenced by the data being used as well as the imagination and technical capabilities of the designer.

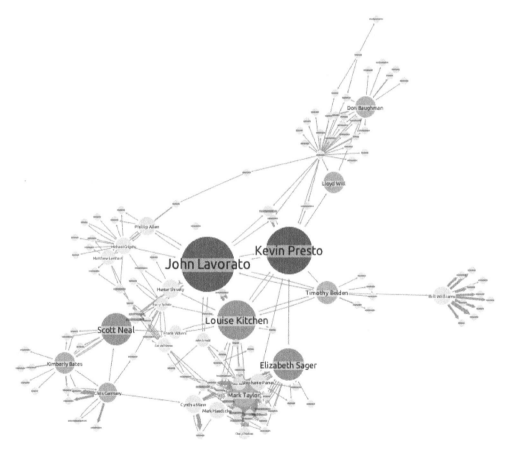

Image created by Joe Parry (http://key-lines.com/)

Mapping geo-spatial data

This final set of examples explores the different ways of mapping geo-spatial data

Choropleth map

Data variables: 2 x quantitative-interval, 1 x quantitative-ratio.

Visual variables: Position, color-saturation/lightness.

Description: As described in the previous chapter, choropleth maps color the constituent geographic units (such as states or counties) based on quantitative values using a sequential or diverging scheme of saturation/lightness. While these are popular techniques, there is a recognized shortcoming caused by the fact that populations are not uniformly distributed. There is a potential distorting effect created by the prominence of larger geographic areas which may not be proportionately representative of the population of data. Make sure you choose your color classifications carefully to ensure you accurately represent the chronological prominence of increasing quantities. An example of a choropleth map is shown in the following screenshot:

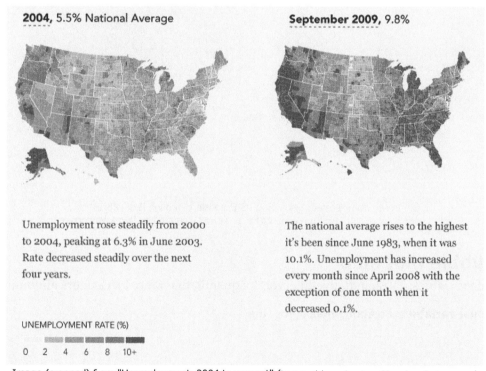

2004, 5.5% National Average

September 2009, 9.8%

Unemployment rose steadily from 2000 to 2004, peaking at 6.3% in June 2003. Rate decreased steadily over the next four years.

The national average rises to the highest it's been since June 1983, when it was 10.1%. Unemployment has increased every month since April 2008 with the exception of one month when it decreased 0.1%.

UNEMPLOYMENT RATE (%)

0   2   4   6   8   10+

Image (cropped) from "Unemployment, 2004 to present" (http://projects.flowingdata.com/america/unemployment/raw.html), created by Nathan Yau

Dot plot map

Data variables: 2 x quantitative-interval.

Visual variables: Position.

Description: A dot plot map essentially displays a geographical scatter plot of records, combining the longitude and latitude to position marks on the map. In the following example, we also see this data being gradually plotted over time to reveal a story of geographical spread:

Image from "Posted: Visualizing US Expansion Through Post Offices" (http://blog.dwtkns.com/2011/posted/), created by Derek Watkins

Bubble plot map

Data variables: 2 x quantitative-interval, 1 x quantitative-ratio, 1 x categorical-nominal.

Visual variables: Position, area, color-hue.

Description: This type of mapping plots differently-sized circular markers over given geographical coordinates to indicate the magnitude of a quantitative value. Whereas the dot plot maps were like geographical scatter plots, these are essentially bubble charts overlayed on to a map. The main contention with these designs tend to be that the spread of bubbles, depending on their size, can reach far beyond their geographical point and end up bleeding into other circles. Normally, the colors used include a relatively high transparency setting in order to accommodate the potential overlaps and "halos" are often used to distinguish outer edges. An example of a bubble plot map is shown in the following screenshot:

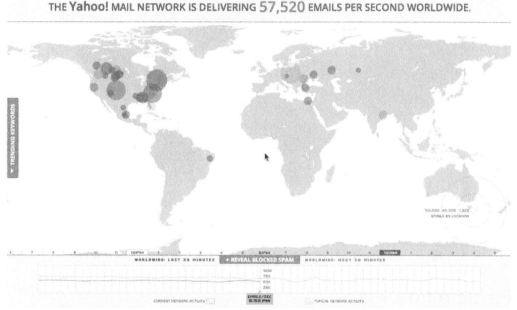

Image from "Visualizing Yahoo! Mail" (`http://www.periscopic.com/#/work/yahoo-mail/`), created by Periscopic

Isarithmic map (or contour map or topological map)

Data variables: Multiple x quantitative, multiple x categorical.

Visual variables: Position, color-hue, color-saturation, color-darkness.

Description: This is a technique for overcoming the flaws associated with the choropleth map and involves combining color-hue (to represent a political party), with color saturation (to represent the dominance of party persuasion), with a final dimension of color-darkness to represent the density of population. Algorithms are applied to help smooth the representation through the contour effect and this creates an elegant end result; as shown here:

Image from "Isarithmic Maps of Public Opinion Data" (http://dsparks.wordpress.com/2011/10/24/isarithmic-maps-of-public-opinion-data/), created by David B Sparks

Particle flow map

Data variables: Multiple x quantitative.

Visual variables: Position, direction, thickness, speed.

Description: A particle flow map uses animation to portray the motion of data across geography and over time. In the following example, we see the motion of the currents that drive the world's oceans. These careful and highly sophisticated constructions combine multiple variables of location, size, speed, and direction to create a compelling design that perfectly captures the nature of the subject matter.

Image from "Perpetual Ocean" (http://www.nasa.gov/topics/earth/features/perpetual-ocean.html), created by NASA/Goddard Space Flight Center Scientific Visualization Studio

Cartogram

Data variables: 2 x quantitative-interval, 1 x quantitative-ratio.

Visual variables: Position, size.

Description: Where a choropleth map takes a location and gives it a shade of color to represent a value, a cartogram takes a location and resizes the geographic shape to represent a value. The result is a distorted and skewed view of reality in the form of a reconfigured atlas. As with many of the chart types outlined here, the purpose is not to enable exact readings, rather to highlight the highly inflated, deflated, and unchanged shapes and sizes. They do rely on a certain predeveloped familiarity of (for example) a country's position, its shape, and its size. The most effective deployment of such charts tends to be when they are interactive and you can unlock all the benefits of exploratory analysis. An example is given here:

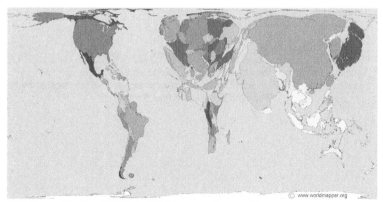

Image from "All Cancer Deaths: 2002" via Worldmapper (http://www.worldmapper.org/images/)
©Copyright SASI Group (University of Sheffield)

Dorling cartogram

Data variables: 2 x categorical, 1 x quantitative-ratio.

Visual variables: Position, size, color-hue.

Description: A Dorling cartogram (named after Professor Danny Dorling who invented them) deploys a uniform shape (typically a circle) to represent a geographical location and then sizes this according to a quantitative variable. In the following example, we see a portrayal of countries represented by circles, sized according to that country's CO2 emissions and colored to distinguish the continents. As before, we may struggle to easily identify places that have now been transformed in shape, size, and position but effective annotation can generally compensate for that.

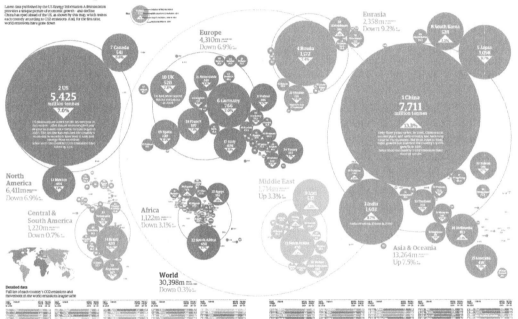

Image from "An atlas of pollution: the world in carbon dioxide emissions" (`http://image.guardian.co.uk/sys-files/Guardian/documents/2011/02/10/CarbonWeb.pdf`), created by Michael Robinson, Mark McCormick and Paul Scruton

Network connection map

Data variables: 2 x quantitative-interval, 1 x categorical-nominal.

Visual variables: Position, link, color-hue.

Description: Similar to the network diagram that we saw before, the intention of a network connection map is to facilitate the exploration of complex geographical connections. A connection map joins up related locations to form a pattern that enables discovery of hubs, overlaps, clusters, and gaps—pretty much the same focus as that of the network diagram but this time with the platform of geographical coordinates. The exhaustiveness of certain datasets means an image of the atlas is almost fully formed from the resulting patterns and there is no need for the actual mapping layer to be visible. In the following image, we see the world's flight paths with those involving Toronto highlighted in orange.

Image from "Toronto Flight Lines" (http://www.biodiaspora.com/), created by Bio.Diaspora 2012

Summary

This chapter has showcased a fairly comprehensive range of visualization chart types across a taxonomy of different methods.

These examples should be considered as representative of the majority of the relevant and popular approaches being used today. However, you should appreciate that this is not an exhaustive nor restrictive collection of options and does not cover all the derivatives that are possible.

The purpose of this chapter was to help you to understand the challenges and options involved in rationalizing your data representation solutions. Hopefully, the information provided about the relationship between the physicality of your data variables and the chart types that can accommodate them will give you a fast track solution.

Additionally, you may have learnt more about the different roles played by visual variables across these examples, and it may inspire you to have the confidence to consider constructing your own unique solutions.

From what we have covered in this chapter and the previous chapter, you should be fairly clear about the design choices you are going to make. We now need to move beyond concept and towards production.

In *Chapter 6, Constructing, Launching, and Evaluating the Data Visualization*, we take the methodology through to the finish with a discussion about some of the most important tools and resources available for creating and launching visualizations. We will then discuss evaluation and the ongoing challenge of developing your visualization design capabilities.

6

Constructing and Evaluating Your Design Solution

The work we have undertaken over the past two chapters has helped us to shape and refine our design concept leading to a visual specification that we believe will most effectively deliver against the requirements of our project. This completes our preparation work and we now move away from concepting and towards construction.

In this final chapter — and last stage of the methodology — we look at the broad variety of options for building our solution and the remaining important tasks to undertake before launching.

We will run through a selection of the most common and useful software applications and programming environments to help you select the most appropriate tool to match your design requirements and technical capabilities.

We will look at some of the key considerations around testing, finishing, and launching a design solution as well as the important matter of evaluating the success of your project post-launch.

Finally, we wrap things up with a discussion about the best ways for you to continue to learn, develop, and refine your data visualization design skills as you seek to master this fascinating and rewarding discipline.

For constructing visualizations, technology matters

The importance of being able to rationalize options has been a central theme of this book. As we reach the final stage of this journey and we are faced with the challenge of building our visualization solution, the keyword is, once again, choice.

The intention of this book has been to focus on offering a handy strategy to help you work through the many design issues and decisions you're faced with.

Up to now discussions about issues relating to technology and technical capability have been kept to a minimum in order to elevate the importance of the preparatory and conceptual stages. You have to work through these challenges regardless of what tools or skills you have.

However, it is fair to say that to truly master data visualization design, it is inevitable that you will need to achieve technical literacy across a number of different applications and environments.

All advanced designers need to be able to rely on a symphony of different tools and capabilities for gathering data, handling, and analyzing it before presenting, and launching the visual design. While we may have great concepts and impressively creative ideas, without the means to convert these into built solutions they will ultimately remain unrealized. The following example, tracking 61 years of tornado activity in the US, demonstrates a project that would have involved a great amount of different analytical and design-based technical skills and would not have been possible without these:

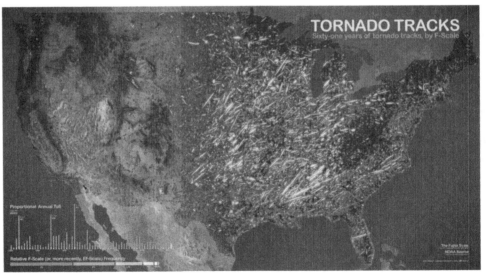

Image from "Tornado tracks" (http://uxblog.idvsolutions.com/2012/05/tornado-tracks.html), created by John Nelson/IDV Solutions.

In contrast to most of the steps that we have covered this far, the choices we make when it comes to producing the final data visualization design are more heavily influenced by capability and access to resources than necessarily the suitability of a given tool. This is something we covered earlier when identifying the key factors that shape what may or may not be possible to achieve.

To many, the technology side of data visualization can be quite an overwhelming prospect—trying to harness and master the many different options available, knowing each one's relative strengths and weaknesses, identifying specific function and purpose, keeping on top of the latest developments and trends, and so on.

Acquiring a broad technical skillset is clearly not easily accomplished. We touched on the different capability requirements of data visualization in *Chapter 2, Setting the Purpose and Identifying Key Factors*, in the *The "eight hats" of data visualization design* section. This highlighted the importance of recognizing your strengths and weaknesses and where your skillset marries up with the varied and numerous demands of visualization design. In order to accommodate the absence of technical skills, in particular, you may need to find a way to collaborate with others or possibly scale down the level of your ambition.

Visualization software, applications, and programs

The scope of this book does not lend itself to provide a detailed dissection and evaluation of the many different possible tools and resources available for data visualization design. There are so many to choose from and it is a constantly evolving landscape—it feels like each new month sees an additional resource entering the fray. To help, you can find an up-to-date, curated list of the many technology options in this field by visiting `http://www.visualisingdata.com/index.php/resources/`.

Unlike other design disciplines, there is no single killer tool that does everything. To accommodate the agility of different technical solutions required in this field we have to be prepared to develop a portfolio of capabilities.

What follows is a selection of just some of the most common, most useful, and most accessible options for you to consider utilizing and developing experience with. The tools presented have been classified to help you understand their primary purpose or function.

Charting and statistical analysis tools

This category covers some of the main charting productivity tools and the more effective visual analytics or Business Intelligence (BI) applications that offer powerful visualization capabilities.

Microsoft Excel (`http://office.microsoft.com/en-gb/excel/`) is ubiquitous and has been a staple diet for many of us number crunchers for most of our working lives. Within the data visualization world, Excel's charting capabilities are somewhat derided largely down to the terrible default settings and the range of bad-practice charting functions it enables. (3D cone charts, anyone? No, thank you.)

However, Excel does allow you to do much more than you would expect and, when fully exploited, it can prove to be quite a valuable ally. With experience and know-how, you can control and refine many chart properties and you will find that most of your basic charting requirements are met, certainly those that you might associate more with a pragmatic or analytical tone.

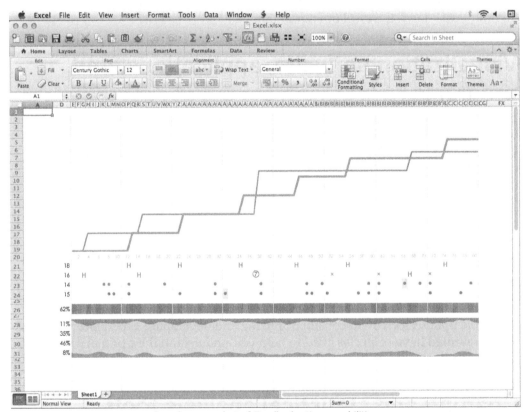

Sample screenshot of Excel's charting capabilities

Excel can also be used to serve up chart images for exporting to other applications (such as Illustrator, see later). Search online for the work of Jorge Camoes (`http://www.excelcharts.com/blog/`), Jon Peltier (`http://peltiertech.com/`), and Chandoo (`http://chandoo.org/`) and you'll find some excellent visualization examples produced in Excel.

Tableau (`http://www.tableausoftware.com/`) is a very powerful and rapid visual analytics application that allows you to potentially connect up millions of records from a range of origins and formats. From there you can quickly construct good practice charts and dashboards to visually explore and present your data. It is available as a licensed desktop or server version as well as a free-to-use public version.

Tableau is particularly valuable when it comes to the important stage of data familiarization. When you want to quickly discover the properties, the shapes and quality of your data, Tableau is a great solution. It also enables you to create embeddable interactive visualizations and, like Excel, lets you export charts as images for use in other applications.

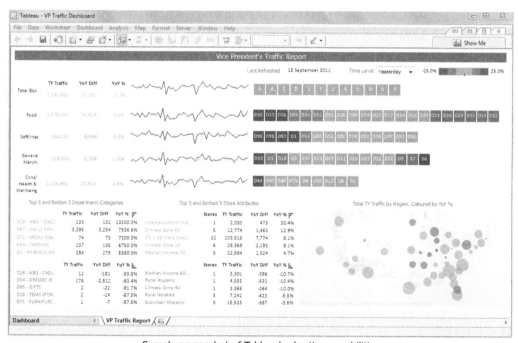

Sample screenshot of Tableau's charting capabilities

There are many excellent Tableau practitioners out there whose work you should check out, such as Craig Bloodworth (`http://www.theinformationlab.co.uk/blog/`), Jérôme Cukier (`http://www.jeromecukier.net/`), and Ben Jones (`http://dataremixed.com/`), among many others.

While the overall landscape of BI is patchy in terms of its visualization quality, you will find some good additional solutions such as QlikView (`http://www.qlikview.com/uk`), **TIBCO Spotfire** (`http://spotfire.tibco.com/`), **Grapheur** (`http://grapheur.com/`), and **Panopticon** (`http://www.panopticon.com/`).

You will also find that there are many chart production tools available online. Google has created a number of different ways to create visualizations through its **Chart Tools** (`https://developers.google.com/chart/`) and **Visualization API** (`https://developers.google.com/chart/interactive/docs/reference`) environments. While you can exploit these tools without the need for programming skills, the API platforms do enable developers to enhance the functional and design options themselves.

Additionally, **Google Fusion Tables** (`http://www.google.com/drive/start/apps.html`) offers a convenient method for publishing simple choropleth maps, timelines, and a variety of reasonably interactive charts.

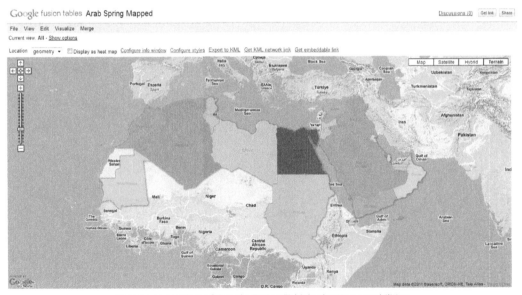

Sample screenshot of Google Fusion Table's charting capabilities

Other notable browser-based tools for analyzing data and creating embeddable or exportable data visualizations include **DataWrapper** (`http://datawrapper.de/`) and **Polychart** (`http://polychart.com/`).

One of the first online offerings was Many Eyes, created by the IBM Visual Communications Lab in 2007, though ongoing support and development has lapsed. Many Eyes introduced many to **Wordle** (`http://www.wordle.net/`) a popular tool for visualizing the frequency of words used in text via "word clouds". Note, however, the novelty of this type of display has long since worn off for many people (especially please stop using it as your final PowerPoint slide in presentations!).

Programming environments

The ultimate capability in visualization design is to have complete control over the characteristics and behavior of every mark, property, and user-driven event on a chart or graph. The only way to fundamentally achieve this level of creative control is through the command of one or a range of programming languages.

Until recent times one of the most important and popular options was **Adobe Flash** (`http://www.adobe.com/uk/products/flash.html`), a powerful and creative environment for animated and multimedia designs. Flash was behind many prominent interactive visualization designs in the field. However, Apple's decision to not support Flash on its mobile platforms effectively signaled the beginning of the end. Subsequently, most contemporary visualization programmers are focusing their developments on a range of powerful JavaScript environments and libraries.

D3.js (`http://d3js.org/`) is the newest and coolest kid in town. Launched in 2011 from the Stanford Visualization Group that previously brought us Protovis (no longer in active development) this is a JavaScript library that has rapidly evolved into to *the* major player in interactive visualization terms.

D3 enables you to take full creative control over your entire visualization design (all data representation and presentation features) to create incredibly smooth, expressive, and immersive interactive visualizations. Mike Bostock, the key creative force behind D3 and who now works at the New York Times, has an incredible portfolio of examples (http://bost.ocks.org/mike/) and you should also take a look at the work and tutorials provided by another D3 "hero", Scott Murray (http://alignedleft.com/).

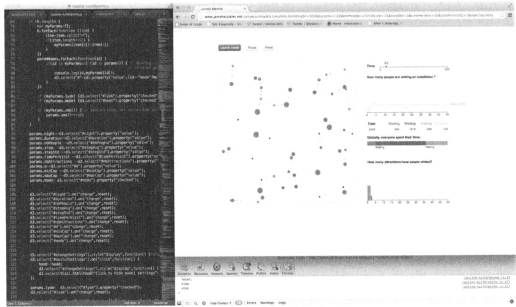

Sample screenshot of D3.js development environment

D3 and Flash are particularly popular (or have been popular, in the latter's case) because they are suitable for creating interactive projects to work in the browser.

Over the past decade, **Processing** (http://processing.org/) has reigned as one of the most important solutions for creating powerful, generative, and animated visualizations that sit outside the browser, either as video, a separate application, or an installation. As an open source language it has built a huge following of creative programmers, designers, and artists look to optimize the potential of data representation and expression. There is a large and dynamic community of experts, authors, and tutorial writers that provide wonderful resources for anyone interested in picking up capabilities in this environment.

There are countless additional JavaScript libraries and plugins that offer specialist capability, such as **Paper.js** (http://paperjs.org/) and **Raphaël** (http://raphaeljs.com/), to really maximize your programming opportunities.

Moving briefly away from interactive programming environments we turn to **R** (`http://www.r-project.org/`), a highly extensible, open source language for statistical analysis and graphical techniques. R has developed into a powerful and versatile method for creating static charts and graphics that transcend the creative limitations of software packages such as Excel. There is a large active online community, which can really help with the challenge of going through the learning process. To demonstrate R's worth, the New York Times use this extensively in their data sketching and static graphic workflows. Check out **Mondrian** (`http://rosuda.org/mondrian/`) and **Wolfram Mathematica** (`http://www.wolfram.com/mathematica/`) for other powerful, statistical graphing capabilities.

Quadrigram (`http://www.quadrigram.com/`) is an innovative visual programming environment designed to enable anyone working with data to create powerful, flexible, and custom visualizations. It is intended to be accessible for people with limited technical and programming experience.

In support of open interactive journalism, the **Miso** project (`http://misoproject.com/`) developed by the Guardian and Bocoup, is a fairly recent arrival. It provides open source tools for developers and non-developers alike to facilitate the quick creation of impressive, extensible interactive data visualizations. There also exists an option for developers to get under the hood and extend and expand the computational methods.

Other notable programming tools to mention include **Nodebox** (`http://nodebox.net/`), which is a Python-based tool for creating generative, static, animated, or interactive visualizations, and **KendoUI** (`http://www.kendoui.com/`) for building interactive HTML5-based data visualizations for both web and mobile applications.

Finally, in recent times, we have witnessed the rise of **WebGL** (`http://www.chromeexperiments.com/webgl/`), a new web technology for rendering interactive two-and three-dimensional graphics. The utilization of this standard has so far seen more experimentation than particularly solid visualization work, but it certainly offers new capabilities for pushing the creative boundaries of data representation.

Tools for mapping

The great opportunity that exists these days for plotting geo-spatial data onto maps is matched by the range of tools available to accomplish it.

Powerful options come in different shapes and forms through **Arc GIS** (http://www.esri.com/software/arcgis), **Indiemapper** (http://indiemapper.com/), **Instant Atlas** (http://communities.instantatlas.com/), **Geocommons** (http://geocommons.com/), and **CartoDB** (http://cartodb.com/). Across these options you will find the ability to create, rich interactive visualizations of geo-spatial data and full-on mapping applications, typically offering flexible licensing and pricing plans from free/trial through to premium levels depending on your needs.

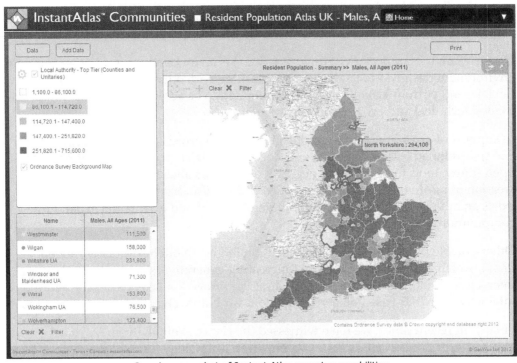

Sample screenshot of Instant Atlas mapping capabilities

For those designers with developer skills wishing to have greater creative control and freedom over their mapping solutions, there are a number of open source mapping frameworks and libraries such as **Polymaps** (http://polymaps.org/), **Kartograph** (http://kartograph.org/), **Leaflet** (http://leafletjs.com/), and **OpenStreetMap** (http://www.openstreetmap.org/).

Additionally, **TileMill** (http://mapbox.com/tilemill/) offers an extremely versatile and accessible application for making elegant data-driven maps whether you are a beginner designer or more-established cartographer.

Other specialist tools

Not all visualizations are interactive, of course, and some of the finest visualization works we see are static pieces. Infographics in particular are typically manually crafted designs, comprising a blend of different visual design elements (such as charts, illustrations, and diagrams). As we have already mentioned, often the chart elements we use for our static work originate from tools such as Excel, Tableau, or R with images imported to help construct a final work.

The vast majority of statics are produced using **Adobe Illustrator** (http://www.adobe.com/uk/products/illustrator.html), the long-established and all-powerful creative package that has been *the* graphic and illustration tool for many years. There is now an open source alternative called **Inkscape** (http://inkscape.org/) ,which offers an impressive array of features that offer a viable alternative for many peoples' needs.

For many people (perhaps those with limited access to varied resources) **PowerPoint** (http://office.microsoft.com/en-gb/powerpoint/) or **Keynote** (http://www.apple.com/uk/iwork/keynote/) provide a perfectly adequate platform for their data presentation needs. Another Adobe package, **InDesign** (http://www.adobe.com/uk/products/indesign.html) provides a further means for creating and publishing final static works.

Elsewhere, for network visualizations, exploratory graphs, and analysis of complex systems check out **KeyLines** (http://key-lines.com/) and **Gephi** (https://gephi.org/).

If you're looking to create advanced motion graphics, modeling, simulation, and visual effects then **Maya 3D** (http://usa.autodesk.com/maya/) and **Adobe After Effects** (http://www.adobe.com/uk/products/aftereffects.html) are incredibly powerful, industry standard production platforms.

Finally, to showcase your static work, once you've created your final designs and want to publish and share image files, sites such as **closr.it** (http://www.closr.it/) or **zoom.it** (http://zoom.it/) enable navigable, zoom-able windows to host large, detailed images.

The construction process

So, you've selected the tools you'll need to build your design and you are now well in to the execution stage. We're not far from the finishing line but it's not yet time for you to lower your guard, lose your focus, or cease your momentum.

You see, this is the part of the design process where stresses and strains emerge—the ill-timed bugs, dataset problems, functional failures, unwanted interference. During this stage it is important that you keep your cool and see your tasks through as efficiently as possible.

As you work through the construction process, it is important to focus on getting the functional elements of your solution working first before spending too much time achieving your desired aesthetic or incorporating technical flair. It is always very tempting to spend too much time, too soon on things that shouldn't really be given such priority. Just remind yourself that there is no point running out of time trying to make something look good when it doesn't yet function. Remember, it will be easier to make something that is functional, beautiful, than it is to make something beautiful, functional.

As we mentioned earlier in the book, you will rarely create a worthy project without the need for iteration. While we have to present the sequences of the methodology in this book in linear fashion, there is always going to be movement forward and backwards between stages. This is something that should be accepted but also embraced—it is part and parcel of any creative process. While a methodological approach to this challenge gives you structure and a neat framework of concerns to work through, iteration gives you the creative breathing room to allow different ideas to blossom and influences to take hold. It is something that you should be prepared to do and plan for.

You clearly want to avoid long iterative cycles but smaller ones can really help you explore, clarify, and refine your potential solutions. It may be that you end up following two or three parallel options to quite an advanced stage and then see which emerges as the strongest. Indeed, some clients will state a need for evidence of alternatives before committing themselves. For these client-based projects, you need to maintain open dialog throughout to avoid any inconsistency in interpretation from either party. Do your absolute best to eradicate the possibility of last minute surprises about a solution not matching requirements or expectations. That is a sting in the tail nobody wants!

As you work through your construction stages, in particular, there will be points when you recognize a need to make certain sacrifices. There may be things you intend to include but can't justify them. Trade-offs are a constant necessity caused by time or resource constraints.

Some of the things we find hardest to drop are the most irrational. We often find ourselves in a sense of denial. This may come from a desire to include features that you have slaved over or become overly precious about.

We saw an example of this in the discussion about color in *Chapter 4, Conceiving and Reasoning Visualization Design Options*. Here we saw an initially conceived title format for an Olympics project that was formed out of thumbnail images of all the historical event posters; this image is shown here:

When it came to incorporating this title into the final piece, it was clear that it drew too much attention away from the rest of the visualization. Despite this being obvious, because of the time and energy spent on making this title image, it was hard to relinquish. Thankfully, a sensible voice determined that we should drop it and find a simpler solution. Simple advice? Take the hit and just get over it!

THE PURSUIT OF FASTER
Visualising the evolution of Olympic speed

Image from "Pursuit of Faster" (http://www.visualisingdata.com/index.php/2012/07/new-visualization-design-project-the-pursuit-of-faster/), by Andy Kirk and Andrew Witherley

As we approach the maturing stages of our development work, this idea of getting input from others becomes more important. It can be quite a tough moment to convince yourself that something is ready to be judged (in a prelaunch setting) but it is invaluable to test out people's responses to what you are creating.

You want people who are informed about the context of the work and also about the challenges involved in creating a visualization. You also need to trust them to give you constructive and reasonable feedback, otherwise it may prove a wasted effort.

You should be seeking feedback on a number of dimensions of your design in order to determine if the intention of your solution is consistent with the audience experience:

- What is their instinctive reaction? Positive, negative, intrigue, confusion, or just a plain "so what?"

- Can they understand how to read the graphic or use the tool? Does it have clear explanations and intuitive design in terms of visual hierarchy and structural arrangement?

- Can they derive insight from it? Maybe throw them some test questions to assess the visualization's ability to effectively inform.

- Does it work functionally? Can they find any errors, mistakes, programmatic errors, or any other design flaw that undermines the clarity, accuracy, or performance of the solution?

There are plenty of evaluation methodologies and techniques, probably much more sophisticated than this, but these are just some of the most useful prompts for you to gather feedback against before finalizing your work.

Approaching the finishing line

Here is a quote from Antoine de Saint-Exupery:

> *"You know you've achieved perfection in design, not when you have nothing more to add, but when you have nothing more to take away."*

The finishing line is now getting ever closer. However, apart from those projects where there is a clear finite deadline to work to, the judgment of when a design is actually finished is not necessarily always obviously recognizable. A deadline provides this finality, but open-ended projects need their own completion point to be determined. It is natural to keep tweaking, refining, and enhancing your piece but eventually you need to call out something as being completed.

A useful signpost to note your progress was proposed by designer Martin Wattenberg (co-developer on the "Wind Map" project that we saw earlier). Martin describes the subtle but telling change in your role as you shift from debugging a design (programmatically or figuratively) to finding yourself becoming an enthusiastic user, engaging with your own work to unearth insights.

As the quote at the start of this section expresses, another viewpoint is to step back and away from your design and challenge everything that you have included. Justify to yourself (and/or to others) the reason why features or design choices need to stay, but also determine what elements you can eliminate, those that don't add any communicative or functional value. It's not necessarily about striving for minimalism; rather the most elegant and clear form.

As well as challenging all our design choices we also need to switch mindsets more towards the Project Manager or Administrator's perspective and undertake some important checks. Sometimes, when you're close to finishing you would prefer to stick your head in the oven than seek issues that need addressing, but you've got to continue to strive for optimal accuracy and intercept any potential mistakes.

Simple errors can completely undermine quality—an extra zero in a value, the mislabeling of a country, an emboldened font when it wasn't wanted, and so on. It might be the smallest and most innocent of mistakes, but that can be enough to tarnish the rest of your work with doubt in the eyes of your audience.

Try to see this as the final push. Paying attention to the finer details of your work will safeguard the project's integrity (and by extension your own, as the designer). Hopefully, much of the user-testing and evaluation work outlined just before will help in the identification of any problems in accuracy. People with a freshness of perspective can often provide great value on this front.

Whether it is them or you looking for these characteristics, here are a few things you need to watch out for:

- **Data and statistical accuracy**: Scan through a good-sized sample of all your visualized data values to ensure there aren't any erroneous items or incorrect outliers. Check the rigor of all your statistics and calculations.

- **Visualization accuracy**: Make sure that the way you have represented your data is functioning effectively and does not mislead the user or reader. Do all your representation choices accurately portray the data values they're associated with?

- **Functional accuracy**: More concerned with interactive pieces—do all the functions and features on your design perform as you intended?

- **Visual inference**: As we stated before, visual inference should equal data inference. If it looks like data, it should be data. If something looks significant, maybe through its positioning or color choice, then it should be significant. If there is any decorative element or other artifact that appears to be implying something it is not meant to, remove it.

- **Formatting accuracy**: Check the consistency of your typography, in terms of type, style, and size. Make sure your color usage is consistent down to the RGB or CMYK code level.

- **Annotation accuracy**: Read through all your titles, labels, introductory text, credits, captions, and check any units that you have included. It's not just about spelling or grammatical errors but checking to see if things make sense and are succinctly expressed.

Post-launch evaluation

The exciting and also probably anxious moment has arrived and your visualization has now been launched in to the wild!

How, where, and what this launch actually looks like clearly covers a very broad range of possibilities—it might be a chart in a report, a presentation to a board meeting, an infographic in a newspaper, or an interactive web-based project.

Regardless of how this piece exists, in an ideal world you would now seek to assess the visualization's effectiveness and impact in a post-launch setting. I say in an ideal world because sometimes you simply don't have sufficient capacity or resources to allocate to the post-launch evaluation.

However, you should still care to seek an assessment of how well your project has served its purpose. Has the reaction and consequence of the work been consistent with its intent and reason for being created, as we determined earlier in the process? It is important to recall the following terms of reference because they frame the type of feedback we seek:

- Was there a positive reaction to the piece we created?
- Did it deliver the appropriate tone of voice?
- Did it reach the intended audience type and volume?
- Were users able to effectively consume or discover insights?
- Where we had a set idea of the intended consequences of this work, were they experienced?
- What problems did people experience, if any?

To obtain feedback of this type and breadth we must consider multiple channels. Each of the following options provides an incremental level of evaluative value but consequently also requires a proportionate increase in the amount of time, effort, and probably cost to obtain:

- **Metrics and benchmarks**: For web-based visualizations there are a number of easily obtainable measures to indicate the reach and popularity of your project. The traditional analytic measures for page views, visits, and visitors can now be easily supplemented with social media metrics such as Tweet counts, Facebook likes, Google+ shares, and so on. These are very simple, cheap, and accessible indicators to help you form a basic understanding of your design's utilization. What you need to think about is: what does success look like? What are the relative benchmarks of performance against these measures that will inform your overall satisfaction?

- **Client or customer feedback**: Of course, the most tangible form of feedback for many projects will come from those who have asked or commissioned you (and hopefully paid you) to create the solution. You'll learn in no uncertain terms whether or not what you created fell short, matched, or exceeded their expectations. Sometimes, you have to judge yourself against the requirements outlined to you and not the resulting performance. After all, you can only respond to the brief you were given.

- **Peer review**: Sometimes the most important and constructive evaluation can come from peers, perhaps expert practitioners or thought leaders. In the visualization field, there are many examples of bloggers who will conduct a review and critique of new work. Getting visitor hits is one thing but receiving a positive review and mention from a peer is worth its weight in gold.

- **Unstructured feedback**: This type of evidence might come via online comments forms, reaction on social media, or through anecdotal channels (e-mails, in-person conversations, perhaps overheard comments) to add a layer of qualitative reaction and evidence of success or failure.

- **Invite user assessment**: Rather than placing value on anecdotal or reactionary and opportune feedback, you could be more proactive by offering simple mechanisms for users to provide more structured qualitative responses, perhaps through small-scale questionnaires.

- **Formal case studies**: Taking things to a more advanced level of evaluation (almost academic in its nature), case studies can take many forms using techniques such as interviews, observations, and controlled experiments, where you might set tasks, manipulate conditions, and record responses. These will often be undertaken by an independent observer to offer that degree of integrity.

No matter through which of these methods you obtain your evaluation feedback, you should be prepared for and accept criticism. Of course, in this digital age everyone is a critic — and too often anonymous — but you should always welcome constructive feedback and use this to fuel your development.

Finally, from your personal point of view, how effective did *you* think it went? Your own satisfaction is very important because this is what also drives your future decisions and development. Often we'll know best whether something could be considered to be an effective outcome and a satisfying process. Even if the results are very positive, there may be many things you thought could have gone better:

- Did you accomplish the outcomes you wanted?
- Did you create something you were satisfied with?
- Were you satisfied with how you rationalized the choices?
- Maybe you hated the project, the client, or the subject matter
- Perhaps you spent far too long on the work and you haven't been paid or rewarded sufficiently for the time you invested
- Maybe you regret consuming so much caffeine late at night

Try not to weigh yourself down with too many thoughts of regret around "wish I'd not done this" or "wish I'd managed to include that". Instead, put all reflection to best use as a learning experience to inform your development and preparedness for future opportunities.

Developing your capabilities

The project is over. You can take a deep breath and relax. Well, at least for an hour until your next project is lined up!

For you the bigger picture now is to consider your ongoing development in this discipline, learning from each experience, and building up your expertise.

The single most important message that I want to put across in this book is the value of practice, experience, and ongoing self-improvement. Data visualization is such a multidimensioned and rapidly evolving craft that cannot be mastered overnight.

Earlier we looked at the framework of the "eight hats". Through assessing yourself against this collection of capabilities, skills, and attitudes you can self-determine where your strengths and weaknesses may be and then look to address them. There are several strategies to help ensure your development continues.

Practice, practice, practice!

When it comes to developing your practical design skills the major piece of advice is simple—practice, practice, and more practice. There are so many different variables and subtle challenges involved in every project that you can't fail to learn from each project that you undertake.

We've just reinforced that data visualization is a craft. You need to continue to exercise your creative and analytical muscle to stay in good shape.

If time permits, try forcing yourself to stick to a practice agenda: maybe, do small personal projects every week then a bigger project every month. You might never launch the work in public but just testing yourself against the challenges of gathering data, analyzing, and presenting it will help maintain your development.

An especially ideal opportunity for practice exists through the frequent data visualization contests that are held these days, often with added incentive of prizes for the best-judged work. These typically involve a basic design brief, a published dataset, and a timeframe to create a compelling solution. A great value of these contests comes from seeing all the other solutions that are submitted. This lets you learn how others have tackled the same problem but maybe in different ways as compared to your own approach.

I have already stressed the importance of maintaining a written record of how you have tackled your design projects. It is worth repeating because it will really help you identify areas for improvement both in terms of effectiveness and efficiency. It will also be a useful reference guide should you ever need to take on a similar problem or comparable dataset.

Also, keep all your trash! Whether it is sketches on paper or little developments on the computer that you deemed redundant, where possible, keep them because you never know when they might come in useful.

Earlier in this chapter, we profiled the importance of technology and the potential limitations of your own capabilities. It is up to you to decide how far and the direction you may wish to take your technical skills. You may not always have the time or opportunity but if you are really serious about advancing your visualization design skills, you should try to push yourself beyond your comfort zone. Rather than relying on the same old tools, pushing them to do things that they're not really designed to do; try out new software, applications, and programming environments. Accept that there will be relatively steep learning curves involved but that the rewards could be great.

Evaluating the work of others

One of the most effective ways of sharpening your visualization design "eye" is by evaluating other designers' work. Not necessarily through providing formal feedback, but just testing your reaction and analysis of the designs you see.

Try to take on the dual mindset of a user and of a designer, in order to undertake a forensic assessment of what has been produced and how well it works using the following prompts:

- What one word describes what is your immediate, instinctive reaction? Is it positive or negative sentiment?

- If it is not necessarily an "instant" piece, does it have the qualities of a "slow-burner", seemingly becoming more appealing after a certain duration?

- What purpose do you think the designer had in mind? Does the style and function of the end product match the likely intention?

- We rarely create these pieces in perfect project conditions, so consider what type of inherent factors might have surrounded and influenced this project? Does a sense of sympathy with the possible influencing factors of the design process effect your impression?

- Work through the five design layers and ask yourself how well each has been executed and what improvements could have been made?

Also think about the general design considerations we outlined about creating accessibility through intuitive design, as well as the idea of reward versus effort, and see how these qualities are achieved.

Eventually, with enough practice, you will develop your critical eye and will become much faster, more informed, and fairer in your judgments of other peoples' work. This will be a great way to educate your own design techniques and refine your own style.

Publishing and sharing your output

One of the contemporary ways of developing your capabilities is to publish yourself. A platform such as a blog will create an ideal means of sharing your work and your ideas.

Posting your design work and building up a public portfolio of your projects creates a virtual shop window. You can take advantage of this format and share narratives about your design process, explaining to people how and why you arrived at the various solutions.

Writing articles, publishing critiques of work, and facilitating discussions are also great ways of promoting yourself. It helps you to learn about the subject. As you write about a topic, you are forced into developing a conviction, to structure arguments, and learn about different perspectives. It really does sharpen your views remarkably, even if (initially, at least) the only visitors to your site are your devoted parents, out of duty.

Eventually, through hard work and dedication, you will create an interested audience and this opens up wonderful opportunities for developing connections with other practitioners, creating rich networks around the world, across demographics, and beyond your subject field.

If you don't have the energy, time, or enthusiasm for a commitment such as a blog, then there are plenty of online galleries and communities through which you can share and publicize your work.

Immerse yourself into learning about the field

Over the past few years we have seen a relative explosion in the amount of online content covering the subject of data visualization, infographics, and data-driven journalism. Websites, blogs, designer sites, and online galleries are now bursting at the seams with interesting articles, new tools, latest projects, and endless amount of inspiration. Social media too is a wonderful platform to learn about key opinions and contemporary developments. The visualization field is particularly active on Twitter where you will find a very spirited and positive community.

Immersing yourself in the array of online resources will keep you up-to-date with contemporary developments. The following is a list of just a small selection of some of the best websites that you should visit and keep a track of. They have been loosely organized by their general remit, though many offer a wide variety of value:

Latest projects, trends, articles, announcements, and developments:

- Visualising Data (http://www.visualisingdata.com/)
- Information Aesthetics (http://infosthetics.com/)
- Flowing Data (http://flowingdata.com/)
- DataVisualization.ch (http://datavisualization.ch/)
- Visual.ly (http://blog.visual.ly/)

Discourse around data visualization:

- Perceptual Edge (http://www.perceptualedge.com/blog/)
- The Functional Art (http://www.thefunctionalart.com/)
- Eager Eyes (http://eagereyes.org/)
- Fell In Love With Data (http://fellinlovewithdata.com/)
- Michael Babwahsingh (http://michaelbabwahsingh.com/)

Design narratives, process, and project critique:

- Charts 'n Things (http://chartsnthings.tumblr.com/)
- The Why Axis (http://thewhyaxis.info/)
- Junk Charts (http://junkcharts.typepad.com/)
- Graphic Sociology (http://thesocietypages.org/graphicsociology/)
- National Geographic (http://juanvelascoblog.com/)

Design or technical tutorials, advice, and much more!:

- Scott Murray (http://alignedleft.com/)
- Jérôme Cukier (http://www.jeromecukier.net/)
- Jim Vallandingham (http://vallandingham.me/)
- Gregor Aisch (http://vis4.net/blog/)
- Naomi Robbins, Forbes (http://blogs.forbes.com/naomirobbins/)

Visualization communities, designers, design agencies and general inspiration:

- Visualizing.org (`http://visualizing.org/`)
- Information is Beautiful Awards (`http://www.informationisbeautifulawards.com/`)
- Any New York Times design (via `http://www.nytimes.com/`)
- Guardian datablog (`http://www.guardian.co.uk/news/datablog`)
- Stamen (`http://content.stamen.com/`)
- Pitch Interactive (`http://www.pitchinteractive.com/beta/index.php`)
- Periscopic (`http://www.periscopic.com/`)
- Moritz Stefaner (`http://well-formed-data.net/`)
- Santiago Ortiz (`http://moebio.com/`)
- Tulp Interactive (`http://tulpinteractive.com/`)

It should go without saying that an intimate appreciation of the many books about and around the subject is vital for learning this craft. Of course, you have already shown great wisdom in choosing this book but there are so many fascinating and invaluable titles to choose from. You can find a list of the most influential titles by visiting `http://www.visualisingdata.com/index.php/resources/`.

It is also important to expose yourself to influences from outside the specific boundaries of this field. You can pick up a great deal of inspiration from reading about graphic design, architecture, product design, typography, move-making, video game design, and journalism—all areas from which we can translate, transport ideas, and learn.

Another critical layer of learning comes from the world of academia and the value of keeping abreast of latest research and studies, from which many tools and best practices naturally emerge. The open access movement is gathering pace and making academic literature much more accessible to those not directly affiliated with academic institutions.

Conferences are also naturally a great way to keep in touch with the very latest developments, hearing from great speakers, and seeing inspirational presentations of case studies and examples. You also get to interact with other similarly passionate practitioners, something that can prove very rewarding.

Beyond these options, clearly a further avenue is through formal training and these days this comes in all shapes and sizes—from online tutorials, video tutorials, and webinars, through to in-person training courses and undergraduate or postgraduate programs. Scour the Internet to find the right solution for you.

Whichever way you go about developing your skills and knowledge, you can be sure there will be plenty of support from across the field. Data visualization is blessed with a wonderful positive and supportive community of incredibly talented and humble people, so you will always be met with a warm welcome.

Summary

We have now come to the end of this design journey. Hopefully, you got to your destination smoothly and you didn't experience many frights along the way!

In this chapter, we have focused on the execution stage of the visualization design process, bringing form to all your preparatory efforts, and transforming your concept into a produced work. We have introduced some of the most useful technologies to give you a flavor of the variety of tools being used for the different stages or types of visualization design.

We have looked at some of the important final steps to take before launching your design, the importance of running final checks across all elements, and conducting testing to get an evaluation of your solution before launch.

Once launched, it is then your prerogative to seek evidence of the impact of your work and we outlined a number of different tactics for undertaking this.

Finally, we suggested some strategies for you to consider pursuing to continue to develop your data visualization skills, knowledge, and experience. This will give you the best chance of taking your capabilities forward and achieving success in this thoroughly exciting field.

Good luck to all of you with the visualization challenges you take on in the future and thank you so much for taking the time to read this book. I hope it helps in any way possible!

Index

Q

QlikView
  URL 164
Quadrigram
  about 167
  URL 167

R

R
  about 167
  URL 167
radial chart
  about 126
  data variables 126
  visual variables 126
radial network
  about 148
  data variables 148
  visual variables 148
Raphaël
  URL 166

S

Sankey diagram
  about 34, 108, 128
  data variables 128
  visual variables 128
scatter plot
  about 144
  data variables 144
  visual variables 144
scatter plot matrix
  about 146
  data variables 146
  visual variables 146
scatterplot matrix visualization 36
sketching 82, 83
skills, data visualization 18, 19
slopegraph
  about 126
  data variables 126
  visual variables 126

sparklines
  about 138
  data variables 138
  visual variables 138
specialist tools, data visualization 169
square pie
  about 133
  data variables 133
  visual variables 133
stacked area chart
  about 140
  data variables 140
  visual variables 140
stacked bar chart
  about 132
  data variables 132
  visual variables 132
stacked column chart. *See* stacked bar chart
statistical analysis tools,
      data visualization 161-164
steam graph
  about 141
  visual variables 141
stories
  finding, example 71-77
  finding, visual analysis used 70, 71
  telling, example 71-77
stream graph
  data variables 141

T

Tableau 59
  about 163
  practitioners 164
  URL 163
table chart. *See* slopegraph
The Why Axis
  URL 179
TIBCO Spotfire
  URL 164
TileMill
  URL 168
topological map. *See* choropleth map

Thank you for buying
Data Visualization: a successful design process

About Packt Publishing

Packt, pronounced 'packed', published its first book "*Mastering phpMyAdmin for Effective MySQL Management*" in April 2004 and subsequently continued to specialize in publishing highly focused books on specific technologies and solutions.

Our books and publications share the experiences of your fellow IT professionals in adapting and customizing today's systems, applications, and frameworks. Our solution based books give you the knowledge and power to customize the software and technologies you're using to get the job done. Packt books are more specific and less general than the IT books you have seen in the past. Our unique business model allows us to bring you more focused information, giving you more of what you need to know, and less of what you don't.

Packt is a modern, yet unique publishing company, which focuses on producing quality, cutting-edge books for communities of developers, administrators, and newbies alike. For more information, please visit our website: www.packtpub.com.

Writing for Packt

We welcome all inquiries from people who are interested in authoring. Book proposals should be sent to author@packtpub.com. If your book idea is still at an early stage and you would like to discuss it first before writing a formal book proposal, contact us; one of our commissioning editors will get in touch with you.

We're not just looking for published authors; if you have strong technical skills but no writing experience, our experienced editors can help you develop a writing career, or simply get some additional reward for your expertise.

MATLAB Graphics and Data Visualization Cookbook

ISBN: 978-1-84969-316-5   Paperback: 284 pages

Tell data stories with compelling graphics using this collection of data visualization recipes

1. Collection of data visualization recipes with functionalized versions of common tasks for easy integration into your data analysis workflow

2. Recipes cross-referenced with MATLAB product pages and MATLAB Central File Exchange resources for improved coverage

3. Includes hand created indices to find exactly what you need; such as application driven, or functionality driven solutions

Circos Data Visualization How-to

ISBN: 978-1-84969-440-7   Paperback: 72 pages

Create dynamics data visualizations in the social, physical, and computer sciences with the Circos data visualization program

1. Transform simple tables into engaging diagrams

2. Learn to install Circos on Windows, Linux, and MacOS

3. Create Circos diagrams using ribbons, heatmaps, and other data tracks

Please check **www.PacktPub.com** for information on our titles

HTML5 Graphing and Data Visualization Cookbook

ISBN: 978-1-84969-370-7   Paperback: 344 pages

Learn how to create interactive HTML5 charts and graphs with canvas, JavaScript, and open source tools

1. Build interactive visualizations of data from scratch with integrated animations and events

2. Draw with canvas and other HTML5 elements that improve your ability to draw directly in the browser

3. Work and improve existing third-party charting solutions such as Google Maps

QlikView 11 for Developers

ISBN: 978-1-84968-606-8   Paperback: 534 pages

Develop Business Intelligence applications with QlikView 11

1. Learn to build applications for Business Intelligence while following a practical case -- HighCloud Airlines. Each chapter develops parts of the application and it evolves throughout the book along with your own QlikView skills.

2. The code bundle for each chapter can be accessed on your local machine without having to purchase a QlikView license.

3. The hands-on approach allows you to build a QlikView application that integrates real data from several different sources and presents it in dashboards, analyses and reports.

Please check **www.PacktPub.com** for information on our titles

Lightning Source UK Ltd.
Milton Keynes UK
UKOW07f1823020615

252771UK00004B/269/P